The Emergence of Relativism

T0256298

Debates over relativism are as old as philosophy itself. Since the late nineteenth century, relativism has also been a controversial topic in many of the social and cultural sciences. And yet, relativism has not been a central topic of research in the history of philosophy or the history of the social sciences to date. This collection seeks to remedy this situation by studying the emergence of modern forms of relativism as they unfolded in the German lands during the "long nineteenth century"—from the Enlightenment to National Socialism. It focuses on relativist and anti-relativist ideas and arguments in four contexts: history, science, epistemology, and politics.

The Emergence of Relativism will be of interest to those studying nineteenth- and twentieth-century philosophy, German idealism, and history and philosophy of science, as well as those in related disciplines such as sociology and anthropology.

Martin Kusch is a professor of philosophy at the University of Vienna and the principal investigator of an ERC Advanced Grant project "The Emergence of Relativism" (2014–2019).

Katherina Kinzel is a postdoctoral researcher in the ERC project "The Emergence of Relativism" (2014–2019).

Johannes Steizinger is a postdoctoral researcher in the ERC project "The Emergence of Relativism" (2014–2019).

Niels Wildschut is a PhD student in the ERC project "The Emergence of Relativism" (2014–2019).

The Emergence of Relativism

German Thought from the
Enlightenment to National Socialism

**Edited by Martin Kusch, Katherina Kinzel,
Johannes Steizinger, and Niels Wildschut**

Routledge
Taylor & Francis Group

LONDON AND NEW YORK

First published 2019
by Routledge
4 Park Square, Milton Park, Abingdon, Oxon OX14 4RN
605 Third Avenue, New York, NY 10017

First issued in paperback 2023

Routledge is an imprint of the Taylor & Francis Group, an informa business

British Library Cataloguing-in-Publication Data
A catalogue record for this book is available from the British Library

Library of Congress Cataloging-in-Publication Data
Names: Kusch, Martin, editor.
Title: The emergence of relativism : German thought from the
Enlightenment to national Socialism / edited by Martin Kusch,
Katherina Kinzel, Johannes Steizinger, and Niels Wildschut.
Description: 1 [edition]. | New York : Routledge, 2019. |
Includes bibliographical references and index.
Identifiers: LCCN 2018050746 | ISBN 9781138571877 (hardback : alk. paper) |
ISBN 9780203702475 (e-book)
Subjects: LCSH: Philosophy, German–19th century. |
Philosophy, German–20th century. | Relativity.
Classification: LCC B2615 .E44 2019 | DDC 149–dc23
LC record available at https://lccn.loc.gov/2018050746

ISBN: 978-1-03-257023-5 (pbk)
ISBN: 978-1-138-57187-7 (hbk)
ISBN: 978-0-203-70247-5 (ebk)

DOI: 10.4324/9780203702475

Typeset in Times New Roman
by Newgen Publishing UK

Publisher's Note
The publisher has gone to great lengths to ensure the quality of this reprint but points out that
some imperfections in the original copies may be apparent.

Contents

Notes on contributors

Terrell Carver is a professor of political theory at the University of Bristol, UK. He is a graduate of Columbia University, New York, and Balliol College, Oxford. He has published texts, translations, and commentary on Marx, Engels, and Marxism, most recently a two-volume study of the "German ideology" manuscripts, and *The Cambridge Companion to The Communist Manifesto*. He has also published extensively on other aspects of political philosophy, including feminist theory and discourse analysis, and currently co-edits the journal *Contemporary Political Theory*.

Kristin Gjesdal is an associate professor of philosophy at Temple University and professor II of philosophy at the University of Oslo. She is the author of *Gadamer and The Legacy of German Idealism* (Cambridge University Press, 2009) and *Herder's Hermeneutics: History, Poetry, Enlightenment* (Cambridge University Press, forthcoming) and a number of articles in the areas of aesthetics, hermeneutics, and German Idealism. She is the editor of *Key Debates in Nineteenth Century European Philosophy* (Routledge, 2016) and a co-editor of *The Oxford Handbook to German Philosophy in the Nineteenth Century* (2015) as well as the *Cambridge Companion to Hermeneutics* (2018).

Sacha Golob is a senior lecturer in Philosophy at King's College London and the co-director of the Centre for Philosophy and the Visual Arts (CPVA). Before joining King's, he was a fellow at Peterhouse, Cambridge. He has published extensively on modern French and German Philosophy. He is the author of *Heidegger on Concepts, Freedom and Normativity* (Cambridge University Press, 2014), and the co-editor of *Cambridge History of Moral Philosophy* (Cambridge University Press, 2018). His current research explores models of ethical development and degeneration.

Katherina Kinzel is a postdoctoral researcher in the ERC Project *The Emergence of Relativism* at the University of Vienna. She has published in the areas of philosophy of science, philosophy of history, Neo-Kantianism, and hermeneutics. Her work has appeared in *Studies in History and Philosophy of Science*, *British Journal for the History of Philosophy*, *HOPOS*, and others.

Martin Kusch is a professor of philosophy at the University of Vienna and the principal investigator of an ERC Advanced Grant project "The Emergence of Relativism" (2014–2019). His main research interests lie in the social history of philosophy, social epistemology, the sociology of knowledge, and the philosophy of the social sciences.

Brian Leiter is Karl N. Llewellyn Professor of Jurisprudence and Director of the Center for Law, Philosophy, and Human Values at the University of Chicago. Authored books include *Nietzsche on Morality* (Routledge, 2nd edition, 2015), *Why Tolerate Religion?* (Princeton, 2013), and *Naturalizing Jurisprudence* (Oxford, 2007). Edited books include *The Oxford Handbook of Continental Philosophy* (2007), *Nietzsche and Morality* (Oxford, 2007), and *The Future for Philosophy* (Oxford, 2004).

Samantha Matherne is an assistant professor in the Department of Philosophy at Harvard University. She specializes in Kant, Neo-Kantianism, Phenomenology, and Aesthetics. Her primary research interests lie in exploring the reciprocal relationship between perception and aesthetics. She approaches these issues largely through a historical lens, as they are taken up by Kant, particularly in his theory of imagination, and developed in Post-Kantian traditions in the 19th and 20th centuries, especially in the work of Maurice Merleau-Ponty, Martin Heidegger, and Ernst Cassirer. She is the author of *Cassirer* (2019) for the Routledge Philosophers Series.

Dermot Moran is Joseph Chair in Catholic Philosophy at Boston College. He has published widely on medieval philosophy and contemporary Continental philosophy, especially the phenomenological tradition. His books include: *The Philosophy of John Scottus Eriugena. A Study of Idealism in the Middle Ages* (Cambridge University Press, 1989; reissued 2004), *Introduction to Phenomenology* (Routledge, 2000), *Edmund Husserl. Founder of Phenomenology* (Cambridge: Polity, 2005), *Husserl's Crisis of the European Sciences: An Introduction* (Cambridge University Press, 2012), and *The Husserl Dictionary* (Bloomsbury, 2012), co-authored with Joseph Cohen. He is a founding editor of *The International Journal of Philosophical Studies* (1993).

Lydia Patton is an associate professor of philosophy at Virginia Tech. Her work has appeared in *Synthese, Studies in History and Philosophy of Science, Kant-Studien, Historia Mathematica*, and *The Oxford Handbook of German Philosophy in the Nineteenth Century*. Her co-edited volume with Walter Ott, *Laws of Nature*, is forthcoming from Oxford University Press. She edited *Philosophy, Science, and History: A Guide and Reader* (Routledge), and co-edited, with Benjamin Jantzen and Deborah Mayo, the issue "Ontology and Methodology" in *Synthese*. She is editor-in-chief of *HOPOS: The Journal of the International Society for the History of the Philosophy of Science*.

Vicki A. Spencer is associate professor of political theory in the Department of Politics at the University of Otago, New Zealand. She is the author of *Herder's Political Thought: A Study of Language, Culture, and Community* (University of Toronto Press, 2012) and the co-editor of *Visions of Peace: Asia and the West* with Takashi Shogimen (Ashgate, 2014) and *Disclosures* with Paul Corcoran (Ashgate, 2000). Her research encompasses seventeenth- and eighteenth-century European thought and contemporary political theory with a focus on culture, identity, and the concepts of recognition and toleration. Her most recent book is *Toleration in Comparative Perspective* (Lexington, forthcoming).

Richard Staley once tried as a teenager to locate Albert Einstein's original papers on relativity in the library of Monash University, Melbourne, only realizing with some shock that of course they were written in German when he got to the pages of the *Annalen der Physik*. Having completed an undergraduate degree in the history and philosophy of science at the University of Melbourne and his PhD at the University of Cambridge on the early career of Max Born, he then taught in the University of Wisconsin-Madison, publishing *Einstein's Generation: On the Origins of the Relativity Revolution* with the University of Chicago Press in 2008. His engagement with Ernst Mach owes much to the six-month period he spent as a research scholar at the Deutsches Museum in 2012, and he now teaches at the University of Cambridge, conducting research on the histories of physics, anthropology, climate change and decolonisation.

Johannes Steizinger is a postdoctoral research fellow in the ERC project "The Emergence of Relativism" at the University of Vienna. His work examines the history of the philosophy of life from Schopenhauer and Nietzsche to National Socialism. He is the author of *Revolte, Eros und Sprache. Walter Benjamins* Metaphysik der Jugend (Berlin, 2013). His articles have appeared in *Politics, Religion & Ideology*, *Studia Philosophica*, and *Benjamin-Studien.*

Niels Wildschut studied philosophy and history in Utrecht and Munich. He currently works in the ERC Project "The Emergence of Relativism" at the University of Vienna, where he is writing his PhD thesis on Johann Gottfried Herder's place in the history of historicism. He has published in *Herder Jahrbuch* and in *Herder on Sympathy and Empathy: Theory and Practice* (Brill, forthcoming).

Paul Ziche studied philosophy, physics, and psychology in Munich at Oxford. He is a professor of the History of Modern Philosophy at Utrecht University. He has published widely on German idealism, in particular the philosophy of Schelling, and on the relationship between philosophy and the sciences in German debates around 1900. He is the author of *Wissenschaftslandschaften um 1900* (Chronos, 2008).

General introduction

Relativism—whether as threat or panacea—has frequently galvanized debate in philosophy, the sciences, and society at large. It was a central topic throughout the twentieth century, and the discussion continues unabated in the twenty-first. Problems of relativism regularly take center stage in arguments over multiculturalism, globalization, political or religious disagreements and their resolution, information bubbles, identity politics, or the authority of science. Debates over relativism extend to all areas of philosophy, including epistemology, ethics, political philosophy, and philosophy of science.

Even though relativism plays a central role in many different discourses, there is little agreement on what the relativist is committing to. Still, the following rough characterization may be useful as guidepost. The relativist holds that judgments or beliefs in a given domain are true or false, justified or unjustified, *only relative to sets of standards*. For the relativist, there is more than one such set, and there is *no neutral way* to rank such sets or to adjudicate between or among them. Some relativists go further and claim that all such sets are *equally valid*.

Relativism can be global or local. Global forms of relativism are meant to apply in a domain-general way. Local forms of relativism are restricted to particular domains, (say, religion, aesthetics, or moral life). Different local forms of relativism result from relativizing to different types of standards. The epistemic relativist relativizes epistemic justification; the moral relativist moral status; the aesthetic relativist aesthetic properties. Forms of relativism also differ in what they regard as the relevant loci of sets of standards; possible loci are, for example, individuals, groups, cultures, paradigms, and genders.

Despite the pivotal role of relativism in debates throughout the twentieth and twenty-first centuries, its history has not to date received the attention it deserves. This lacuna becomes visible when we compare the scholarship on relativism with the scholarship on scepticism. There exist a large number of outstanding papers and monographs on the history of scepticism from antiquity to the early modern period. The historical work has also influenced contemporary work in that some ancient positions have recently been revived.

The situation is rather different in the case of relativism. Important anthologies on relativism (e.g., Blackwell's *A Companion to Relativism* (2011)) contain only scant historical material. There are no monographs tracking the development of relativism over the *longue durée*, and there are few pockets of specialists concerned with relativist themes in particular philosophers. (The exception that proves the rule is the important literature on Plato's *Theaetetus.*) Moreover, when historical forms of relativism are discussed at all, it is usually as predecessors to contemporary views.

This volume constitutes a modest attempt to help remedy the relative neglect of relativism in historical-philosophical scholarship. It focuses on a period that was crucial for the emergence of modern-day forms of relativism: German-language philosophical and scientific discourse of the long-nineteenth-century (roughly from 1770s to the 1930s) and some of its aftermaths. This context was decisive in shaping much of our present understanding of relativism.

Although the term "relativism" gained currency only by the end of the nineteenth century, much of its eventual content was assembled in earlier reflections on how developments in the sciences, the arts, and society had undermined old certainties regarding truth, knowledge, and morality. Relativism emerged as a philosophical view inseparable from a sense of acute intellectual and social crisis. It often acted as something of a covering term for widely perceived intellectual ills such as *psychologism, historicism, sociologism, scepticism, nihilism, subjectivism, and pessimism.*

This anthology explores the German debates over relativism as these unfolded from Herder's anthropology of culture to Gadamer's hermeneutics. It concentrates on four contexts that are of particular importance for the emergence and development of relativism: history, epistemology, science, and politics.

The contributions to this volume seek to illuminate the conceptual and social conditions in which relativistic (or anti-relativistic) themes emerged. Our authors follow the complex historical trajectories of various relativistic themes and arguments in different contexts and subfields of philosophy. In so doing, they make us aware of a wide range of possible attitudes and responses to relativism and help clarify connections between relativism and other *isms*, such as scepticism or nihilism. In this way, they make possible a deeper and more nuanced grasp of why relativism is experienced as threatening and why it continues to be the focal point of broader political and societal debates.

Needless to say, this anthology does not close the gap in our understanding of the history of relativism in the long nineteenth century. We shall be satisfied if it stimulates other historians and sociologists of philosophy to go further and deeper into this fascinating period and topic.

Work on this anthology was made possible by an ERC Advanced Grant (#339382) on "The Emergence of Relativism" located at the Department of Philosophy of the University of Vienna. We are grateful to the other ERC team members (Natalie Ashton, Robin McKenna, Katharina Sodoma) for their support. We also wish to thank Olga Bauer, Karoline Paier, and Ernestine Umscheider for their help. We owe a special thanks to Tony Bruce for accepting this project for publication with Routledge. Our greatest debt is naturally to our contributors for their insightful chapters, their patience, and their enthusiasm.

Vienna, September 2018
The Editors

Part I

History

Introduction

Katherina Kinzel

One of the main tensions that structured historical thought from the eighteenth century onward was that between the universal and the particular. Many scholars in history, philology, and philosophy accepted the Herderian insight that every historical culture is to be viewed as an "individual"— unique, unrepeatable, and centered around values and experiences that may be drastically different from those embraced at present. However, this theme was rarely taken to imply historical relativism. "Individualism" about historical cultures was usually accompanied by a general commitment to the main tenets of "universal history," that is, to the idea that the history of humankind is a unified process with a general spiritual meaning.

Johann Gottfried Herder (1744–1803) himself rejects abstract forms of universalism that detach universal norms from feelings and concrete historical forms of life. He also criticizes the conception of linear historical progress. In his view, the idea that different historical cultures can be ranked hierarchically expresses a self-serving triumphalism that uncritically projects present-day standards onto the past. But while Herder's recognition of radical "historical difference" may seem to pull in the direction of relativism, Herder also maintains that his pluralist vision of historical cultures is compatible with universalism. He articulates his universalism both in reflections on a common human nature and on history as the realization of God's plan in the world.

Although Georg Wilhelm Friedrich Hegel's (1770–1831) systematic speculative philosophy may seem diametrically opposed to Herder's multifaceted and often meandering reflections, Hegel's philosophy of history, too, is shaped by the struggle to reconcile the particular and the universal. Like Herder, Hegel is seeking to show how the seemingly abstract norms of reason acquire concrete forms in the historical process. In order to accomplish this task, he reconceptualizes the tension between the universal and the particular in terms of a relation of *expression* that pertains between the absolute and the relative. In Hegel, the ever-changing *Volksgeist* (national spirit) is the relative expression of the *Weltgeist* (world-spirit), which, in turn, is the historical manifestation of the absolute spirit's movement toward self-consciousness. Of course, Hegel thinks of this process in progressive terms: history has a general meaning, because the "stages" of historical development are also the "stages" of the self-development of spirit toward freedom, self-consciousness and self-identity.

The professionalization of history as a *Wissenschaft* in the early nineteenth century went hand in hand with staunch opposition to this "speculative approach." From Leopold Ranke (1795–1886) to Johann Gustav Droysen (1808–1884) and Jacob Burckhardt (1818–1897), historians were united in rejecting what to them seemed like a schematic apriorism that distorted history in order to derive quick philosophical generalizations. Even though most historians did accept that there was a universal history of humankind, they thought that knowledge of this universal history had to emerge as a hard-won result of the rigorous application of source criticism, the meticulous collection of particular facts, and the judicious compilation of individual national histories.

With this newly forming methodological awareness, the question of how to bridge the gap between the present and "historical others" received a new urgency as well. Although formulated in terms of relativism only rarely, understanding the past came to be viewed increasingly as a problem: too vast seemed the plurality of historical cultures, too great the divides between them. Finding solutions to this problem was one of the central strands that united the hermeneutic tradition and thinkers as diverse as Friedrich Schleiermacher (1768–1834), Droysen, Wilhelm Dilthey (1833–1911) and Hans-Georg Gadamer (1900–2002). In Herder, it is primarily the commonality of human nature that allows for the possibility of understanding "historical others." Schleiermacher places a stronger emphasis on the methodological process that moves between the particular and the general, the part and the whole, thus allowing one to approach—yet never fully reach—an understanding of past linguistic expressions. Dilthey combines both Herderian and Schleiermacherian arguments, and in his later work he uncovers inter-subjective understanding as the epistemological basis not just of textual criticism and history, but of all the *Geisteswissenschaften*. A slightly different solution to the problem of understanding can be found in Droysen and Gadamer, who seek the basis for historical understanding not primarily in human commonalities, or in a methodological process that mediates between part and whole, but rather in historical continuity and tradition.

The methodological discussions in history, philology, and hermeneutics were not without repercussions for philosophy. Toward the late nineteenth century, philosophers began to warn of "historicism," which they considered tantamount to relativism and nihilism. Neo-Kantian philosophers such as Wilhelm Windelband (1848–1915) and Heinrich Rickert (1863–1936) and also historically minded thinkers like Dilthey, and Ernst Troeltsch (1865–1923) worried that the thorough historicization of all life and culture would undermine universal and necessary values. An additional problem emerged from the historicization of philosophy itself: as philosophers jettisoned the Hegelian assumption that the historical development of philosophy expressed a systematic schema, they began to worry that philosophical truths, too, might be relative to historical time, nationality, and culture.

The resulting discussions concerning the historicity of values and philosophical knowledge often intersected with debates over *Weltanschauung.* Philosophers wanted to maintain a role for themselves in adjudicating the conflicts between "materialistic," "positivist," religious, and "historicist" worldviews. Historical plurality came to stand in for political plurality. When Troeltsch worries that historicism leads to the "anarchy of values," he does so in a situation in which the "historicization" of philosophy and theology has seemingly rendered these disciplines incapable of defending a consistent and convincing worldview. And yet, many held out hope that history would not just create a problem of relativism, but also hold the key to its solution. As Dilthey famously put it, the hope was that "the knife of historical relativism ... which has cut to pieces all metaphysics and religion" would also "bring about healing."

The contributions of this section explore the relations among history, philosophy, and relativism, as they unfolded from the late eighteenth to the mid twentieth century. Niels Wildschut analyzes the theological foundations of the historical thought of Herder and Ranke. He argues that it is only within their theology of history that Herder and Ranke appreciate the historical, individual, and diverse. But at the same time, their theology of history ensures that Herder and Ranke approach history from a monist starting point and with the intent to establish the ultimate harmony of history.

Katherina Kinzel focuses on how Windelband and Dilthey responded to the historicization of philosophy. She traces how both philosophers sought to fend off historical relativism by defending the idea of an ahistorical and permanent stratum of philosophical thinking. Kinzel shows, however, that this strategy was a failure: although they succeeded in blocking historical versions of relativism, both authors did so at the cost of incurring a relativism *vis-à-vis* philosophical systems.

Kristin Gjesdal revisits Gadamer's reception of the hermeneutic tradition. She analyzes Gadamer's suggested response to what he conceived as the historicist and relativist shortcomings of hermeneutics. Gadamer's alternative approach views the existence of a continuous tradition as fundamental to our self-understanding as well as our understanding of others. Gjesdal criticizes Gadamer's concept of tradition and contrasts it with Herder's more self-critical approach to historical understanding.

1 Hieroglyphic historicism

Herder's and Ranke's theology of history

Niels Wildschut

Introduction

Johann Gottfried Herder (1744–1803) and Leopold Ranke (1795–1886) are often considered to be among the founding fathers of the tradition of German historicism. Authors like Wilhelm Dilthey (1901), Friedrich Meinecke (1936), and Georg Iggers (1968) have identified in Herder and Ranke's thought central ideas that connect them to later historicist positions. For example, for both authors, historical phenomena have intrinsic value due to their *individuality*; the good in the human-historical world is necessarily *plural*; strong normative implications attach to *diversity*; and while all values depend on broad historical processes, it is necessarily illegitimate and suspect to endeavor to determine *a priori* standards of truth and historical progress. The historicist interpretation generally concludes that, because of these claims, Herder and Ranke must have been committed to historical relativism.

In this chapter, I explain why this conclusion is incorrect. Crucial to my argument is Herder and Ranke's "theology of history" (Löwith 1949). In the literature defending historicist interpretations, its systematic importance has not been properly documented. I claim that it was only within their theology of history that Herder and Ranke appreciated the historical, individual, and diverse.[1] Their theology of history ensured that Herder and Ranke approached the diverse and particular from a monist starting point and with the intent to point out the ultimate harmony of history.

In the first two sections, I start from the theological foundations that Herder and Ranke took to confer upon history its intrinsic value and existential significance. In the third section, I sketch how they sought to order historical material in larger harmonious wholes with the aim of providing a structure for universal history. In this way, I aim to establish that Herder and Ranke were not guilty of some of the crudest versions of the "problem of relativism" or the "crisis of historicism." In the final section, I explore in more detail the views of Herder and Ranke on themes that are often considered relativistic. It is indeed remarkable how many of the relativistic themes identified by Maria Baghramian (in the history of philosophy) and Martin Kusch (in contemporary debates) seem present in Herder and Ranke (Baghramian

2004, 50–82; Kusch 2016, 107–108).[2] Nevertheless, I aim to show that even regarding these themes, the positions of Herder and Ranke strongly diverged from later historicist positions. I conclude that Herder and Ranke should not be interpreted as relativists.[3]

Herder on revelation, history, and faith

Herder's *This Too a Philosophy of History for the Formation of Humanity: Contribution to the Many Contributions of the Century* (hereafter *TTPH*) was published in 1774 (PW 272–358). This short treatise, with its fierce polemic against contemporaneous philosophy of history,[4] has often been considered a foundational text of German historicism (e.g., Beiser 2011, 132). In the same year, Herder published two other works, both of a theological character: a collection of letters, *An Prediger* (*To Preachers*) and an exegesis of the Old Testament, *Älteste Urkunde des Menschengeschlechts* (*Oldest Document of the Human Race*) (FHA 9/1, 67–138, FHA 5, 179–660; Menze 2000, 11). In the latter, Herder puts forward the thesis that an original revelation pervades all of creation. The former presents religious faith as a universal psychological capacity and as part of the human condition. Taken together, these texts provide the foundation on which Herder relied in his 1774 attempt at presenting a unified vision of universal history.

In *TTPH*, Herder confronts the epistemological difficulties relating to this attempt (FHA 4, 32–42, 81–89). Furthermore, he relentlessly criticizes the tendency of his Enlightenment contemporaries to see their own historical standpoints as absolutely privileged and to consider their value-terms as universal tools fit for measuring historical progress.[5] In order to properly assess this intervention by Herder into eighteenth-century philosophy of history, it is crucial to also take into sufficient account his own *alternative* to a one-sided progressivism. Herder's alternative is a providential model according to which history is unified as one development. Herder claims that this development should be understood "in a higher sense than people have imagined it," and that it is extremely difficult to pin down (PW 298, FHA 4, 41).[6] Nonetheless, he is confident that history will appear as *"God's course through the nations"*[7] once it is approached from the right perspective (PW 349, FHA 4, 88). For Herder, identifying this perspective is a religious task. In this section, I show how Herder's philosophy of history builds on his theology.

In *Älteste Urkunde*, Herder puts forward the thesis that God is omnipresent in nature and history (FHA 5, 298). Accordingly, he understands positive religions as ways of capturing a primordial revelation from different historical or geographical standpoints (305 ff.; cf. Gaier 1988, 73–74). This revelation is connected to God's act of creation and handed down to humanity in the form of what Herder calls the *Schöpfungshieroglyphe* (hieroglyph of creation). According to Herder, this hieroglyph is God's primordial revelation to humanity in a seven-step symbol or *Denkbild* (FHA 5, 281; cf. Häfner 1995, 216–221). This *Denkbild* communicates God's act of creation

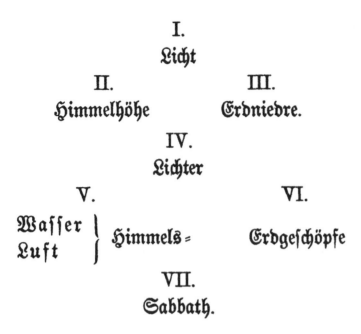

I.
𝔏𝔦𝔠𝔥𝔱

II.
𝔥𝔦𝔪𝔪𝔢𝔩𝔥ö𝔥𝔢

III.
𝔈𝔯𝔟𝔫𝔦𝔢𝔡𝔯𝔢.

IV.
𝔏𝔦𝔠𝔥𝔱𝔢𝔯

V.
𝔚𝔞𝔣𝔣𝔢𝔯 ⎫
𝔏𝔲𝔣𝔱 ⎭ 𝔥𝔦𝔪𝔪𝔢𝔩𝔰 =

VI.
𝔈𝔯𝔟𝔤𝔢𝔣𝔠𝔥ö𝔭𝔣𝔢

VII.
𝔖𝔞𝔟𝔟𝔞𝔱𝔥.

Figure 1 The seven positions of the *Schöpfungshieroglyphe* (Graczyk 2014, 74; Herder 1883, 292, FHA 5, 271).

in a form that prefigures the first written religious documents (FHA 5, 267–282). The hieroglyph structures the seven days of creation in the following way (see fig. 1).

The first book of Genesis is the paradigmatic example of how this *Denkbild* is conserved in narrative form. Positive and natural religion thus coincide. But Herder attempts to establish at length that the hieroglyph informs all other ancient religions as well (305–476). Furthermore, traces of the *Schöpfungshieroglyphe* can be recognized in literary traditions, in nature, and in world history.

One implication of the hieroglyph-theory is that the Biblical period takes center stage as the origin of human culture (FHA 4, 14; Hinrichs 1954, 61). God instructed humanity in its childhood; the hieroglyph's primordial revelation contained the seeds out of which all of humanity's history, culture, and education developed (FHA 5, 297–301, FHA 4, 89). Historical explanation thus exhibits how a thing developed out of its holy origin.[8] This development is dynamic and may present revelation in a variety of shapes: "God is revealed to the human race at many times and in various ways … God's revelations … were *seeds*, concealed and sown in various ways, which contained much that was to develop only with the passing of time and with the extensive passing of time. The Bible consists of the *unfolding of developing times*" (Herder 1993, 213, FHA 9/1, 83). Because of this world-historical significance of the Bible, the *history of religion* becomes Herder's prime pedagogical antidote to theo-logical rationalism (84, 131–132). The Bible's images and examples convey the infinite in the particular, whereas general philosophical concepts merely present abstractions (104).

Another implication of the hieroglyph-theory is that a divine perspective exists in which the *totality* of world history is grasped (FHA 4, 35, 83). In this life, this perspective can never be ours. And God only reveals Himself in forms adjusted to particular historical contexts (18, 48 f.). Furthermore, God does not *intervene* in history directly or supernaturally. God acts through all of nature, and it is the natural development of history that shows that "deity … takes effect so entirely poured forth, uniform, and invisible *through all its works*" (PW 305, FHA 4, 48). Herder's God is immanent in all of nature and history and organizes its interconnectedness.

Due to the presence of God in history, an imaginative and engaged per-spective on all of the experiences of the human soul throughout history can provide us with hints of providence. The senses and feelings may be our guides here: for Herder, the education of the child functions as a model for God's instruction of humanity (FHA 5, 250, FHA 9/1, 102). Faith is related to the senses and to feeling. In Herder's psychology, feeling is a vital aspect of human cognition (FHA 9/1, 84–86, FHA 5, 246–257). Herder thus expounds in *An Prediger* that sensuous faith should be cultured in the right way—through *images* and *history*—rather than dismissed as unphilosophical (FHA 9/1, 100–104, 131; cf. Crowe 2009, 268–269). Furthermore, Herder is convinced that history will appear to the devoted historian as "*the stage of a guiding intention on earth*!, even if we should not be able to see the final intention, the stage of the deity, even if only through *openings* and *ruins of individual scenes*" (PW 299, FHA 4, 42). Herder appreciates the historicity of every individual because he is determined to find hints of this guiding intention in all of his-tory. It is only within this image of creation that individual historical periods have an intrinsic worth. As Herder insists, "Only the *height of revelation offers sight*" (FHA 9/1, 127).

While the difficulty of decipherment is an important connotation of "hiero-glyph," the *Schöpfungshieroglyphe* clearly does not solely refer to the *limits* of human knowledge. In *TTPH*, the seven-step structure of the hieroglyph actu-ally informs the structure of the world-historical development (Pfaff 1983, 409, 415–416). The clearest example is how Herder presents the succession of ancient peoples: the oriental patriarchs occupy the first of the seven positions of the hieroglyph (see fig. 1); the historical development then splits up into an Egyptian (2.) and a Phoenician (3.) position and returns to one position (4.) with the Greek. Herder is less clear, however, how the more recent periods of world history fit into the second run of the hieroglyph. Furthermore, in *TTPH*, Herder refrains from confirming explicitly that the course of world history is hieroglyphic. Apart from the epistemological restrictions on phil-osophy of history, theological limitations also play a role here: the course of history should exhibit its intelligibility by itself; Herder is careful that his theology of history does not appear "fanatic."[9] Still, principally the hiero-glyph provides Herder with a logic for

> *binding together* the most disparate scenes without *confusing* them ... showing how they *relate* to one another, *grow* out of one another, *lose* themselves in one another, all of them taken individually only moments, only through the progression *means to purposes*—what a *sight!*, what a noble *application of human history!*, what *encouragement* to *hope*, to *act*, to *believe*, even where one sees *nothing* or *not everything*.
>
> (PW 299, FHA 4, 42)

These objectives are not merely important for a theory of history. Rather, they directly inform religious practice (65). Furthermore, the seven-step structure of the hieroglyph provides a principle of divine organization that is qualita-tively different from the imposition of abstract categories of moral progress. This is because Herder sees the hieroglyph as a divine *image*; a *Denkbild* for human sensuousness and imagination, which connects different elements into an organic whole (FHA 5, 281, 297–298; Graczyk 2014, 74–76). To sum up, Herder bases his alternative model of history in *TTPH* on a sensualistic conception of faith combined with the view that revelation permeates all of history.

Finally, the hieroglyph-theory also helps to understand Herder's assessment of the present, and thus of the task of the historian. Herder states in *Älteste Urkunde* that on the seventh day, Sabbath, God provided humanity with the education on which all its *Bildung* and happiness depends (FHA 5, 285, 289). The seventh step in *TTPH*'s structure of world history most probably represents Herder's enlightened present (Pfaff 1983, 416). Herder concludes that it is time that humanity turn toward history in order to educate itself and intimate its fate:

> If our age is nobly useful in any respect at all then it is "its *lateness*, its *height*, its *prospect!*" ... Philosopher, if you want to honor and benefit your century's situation—the *book* of *preceding history* lies before you!, *locked* with seven seals, a *miracle book full of prophecy*—the *end of days* has reached you!, read!
>
> (PW 337, FHA 4, 85)

Ranke's theological legitimation of history

Almost sixty years later, Leopold Ranke aimed to liberate the study of history from idealistic philosophy and to legitimate it as an autonomous *Wissenschaft*. He searched for a principle that would explain the tendency of history to oppose the aspirations of philosophy. He summarized it as follows: "History turns with sympathy to the individual ... it likes to attach itself to the conditions of appearance ... it insists on the validity of particular interest. It recognizes the beneficent, the existing, and opposes change which negates the existing. It recognizes even in error its share in truth" (TP 11, AWN 4, 77). Ranke was straightforward about what legitimates this tendency: "It is not necessary for us to prove at length that the eternal dwells in the individual. This is the religious foundation on which our efforts rest. We believe that there is nothing without God, and nothing lives except through God" (TP 11, AWN 4, 77). No further external motivation is necessary for studying history (79). An independent value is ascribed to all historical material.

History can be studied on its own terms because it manifests divine benevolence all by itself (Meinecke 1942, 129–130). The individual appearance has worth because it contains higher principles; the finite because it contains the infinite; the concrete because it contains the abstract (AWN 4, 77). Therefore, universal history is significant not just as the scientific ideal of reaching complete truth but also as a religious accomplishment. The ultimate aim of the historian is to aid humankind in its quest for redemption by uncovering history's hidden meaning (Berding 2005, 43; Gadamer 1960, 215; Hinrichs 1954, 141). Ranke, too, presents this aim by referring to a divine hieroglyph:

> In all of history God dwells, lives and can be recognized. Every deed gives testimony of Him, every moment preaches His name; but most of all ... His presence is plain in the connection of history in the large. God stands there like a holy hieroglyph whose most outer form is apprehended and preserved, perhaps so that He is not lost to more perceptive future centuries ... No matter how it goes and succeeds, let us do our part to unveil this holy hieroglyph! In this way too we serve God, in this way we are also priests and teachers.
>
> (TP 4,[10] Ranke 1949, 18)

For Ranke, any historical appearance has its own relation to the divine. In this way, the infinite can be found in its individual character. Thus, the historical particular is autonomous from philosophical speculation about timeless ideas and linear progress. For Ranke, the infinite is also captured, however, in the connections between all ages. In this case, the individual carries value to the extent that it delivers a contribution to the whole. Hence, the value of the individual is determined only indirectly: the historian attempts to situate the individual in the totality of world history (AWN 4, 297; Krieger 1977, 15–19; Meinecke 1942, 131–132). But this totality itself remains an ideal: "God alone knows world history. We recognize the contradictions—'the harmonies,' as an Indian poet says, 'known to the Gods, but unknown to men'; we can only divine, only approach from a distance" (TP 15, AWN 4, 83).

Ranke's general organization of world history seems structured along a vertical and a horizontal plane. On the vertical plane, there is an immediate relation between God and every individual epoch (TP 2, AWN 4, 60). That is to say that, in every epoch, there is a direct "upward" relation toward

God, and God looks down upon every epoch and sees that it is good. On the horizontal plane, the epochs stand in various relations to each other. Ranke assumes that the succession of historical periods is identical with the divine order of creation (Ranke 1949, 102, 519; Meinecke 1942, 149). Thus, Ranke, too, posits a providential course of world history.

Herder frames the relationship between *individuality* and *development* in a similar way when he states that "no thing in the whole of God's realm … is *only* a means—everything is *means* and *end* simultaneously" (PW 310, FHA 4, 54). Whereas it is legitimate to assess the function of individual periods, all periods preserve an intrinsic purposefulness. Historical periods are situated in immediate relation to God vertically and ordered in a providential sequence horizontally. Furthermore, Herder likewise identifies the providential sequence with the general divine order of creation:

> If the *residential house* reveals a "*divine image*" … how not the *history of its resident*? The former only decoration!, picture in a *single* act, view! The latter an "*endless drama* of *scenes*!, an *epic* of God's through all *millennia, parts of the world*, and *human races, a thousand-formed fable* full of *a great meaning!*"
>
> (PW 336, translation modified, FHA 4, 82–83)

In this endless drama, scenes carry meaning both in themselves and in relation to each other. But in addition, they also all together strive for a final and transcendent end. The individuality of phenomena is always the product of specific historical *developments*. Yet, the notion of development invariably points beyond the individual phenomenon, and allows one to assess its value in terms of its effect, its *contribution* (32). Ultimately, historical significance is defined in terms of outcome (Gadamer 1960, 207). And in their visions of history and salvation, Herder and Ranke formulate expectations regarding the future that lie outside of history (*contra* Koselleck 1975, 674–675; Sikka 2011, 116–125; cf. Löwith 1949, 5–6).

Another way to frame the relation between individuality and development, therefore, is with reference to two onto-theological positions: the panentheism according to which God is present everywhere in creation, and the more traditional Christian position that places salvation (and hence the ultimate meaning of history) beyond history. Both Herder and Ranke relied on versions of panentheism in their arguments for individuality. But regarding the future, the principle of *development* demands a thicker metaphysics: a final end must remain unreached so that the future may preserve a certain promise. This results in a tension between evaluating the particular from the perspective of the divine and from the perspective of a final end. Pointing at the unknown nature of the latter relieves this tension to some extent. In the next section, I sketch some of the concrete structures of world history that Herder and Ranke relied on in their attempts at reconciling these conflicting tendencies.

Individuals and analogies in world history

Previous research has observed that Ranke "applied to history the antinomy characteristic of his romantic age: the idea of an individual reality which was at one and the same time unique in itself and, like other individuals, a manifestation of a universal principle" (Krieger 1977, 16). This "antinomy" had played a major role in German Enlightenment thought ever since Leibniz had put forward his conception of the monad, and already Ernst Cassirer examined how Herder incorporated Leibniz's idea of individuality in his philosophy of history (Cassirer 1916, 116–121). In this section, I will indicate how Herder and Ranke dealt with the antinomy of individual and totality in their accounts of universal history.

Key to their theorizing is that the "universal principle" is not abstract but a holistic whole, a "higher-order individual" (Cassirer 1916, 119; Heinz 1994, 83). This is because the historian's attempt to find unity in history is supposed to be distinct from the philosopher's projection of linear progress onto history. Thus, the universal must not be deduced out of higher principles and projected onto history. Rather, the universal is expressed in the particular, and presents itself to the historian who approaches the particulars of history from the right stance (FHA 4, 32–33; AWN 4, 78, 83, 87–88). Furthermore, the historian, in connecting different ages or nations, must not rely on unifying notions that are specific and local. Instead, the interconnectedness and meaning of world history is *revealed* to the historian by pious faith and divination. It follows that the organization of history can merely be *indicated* by analogies and not be demonstrated from principles. After all, God is present in history (only) in the form of a hieroglyph. Herder and Ranke both attempted to intimate God's plan and stressed its ultimate indiscernibility.

Only by analogies can the *direction* of history's providential development be hypothetically identified. Both Herder and Ranke use the image of a river to convey that the course of history does not depend on predetermined principles; history is "moving more like a river which in its own way determines its course" (TP 22, AWN 2, 62). Other aspects that play a role in this analogy involve the source and the mouth of the river of history. Ranke, for instance, assumes that this river flows from an "active source of life," an "existence kindred to God," and that it searches for "the world-sea" (Ranke 1949, 110). Herder describes "how it sprang forth from a little source, grows, breaks off there, begins here, ever meanders, and bores further and deeper—but always remains *water!, river current!*, drop always only drop, until it plunges into the ocean" (PW 299, FHA 4, 41). In this depiction, too, the image of the river suggests that history started with a *Fall* from a source and is ultimately directed at salvation. Of course, this future "end" is far removed from the present and closed off from philosophical speculation (101). But it does not leave the future entirely *open* either.

According to the tree analogy, history grows out of a seed which, after falling in the earth, first develops in its depths (11). In Herder's *TTPH*, ancient history proceeds from Patriarchal roots along an Egyptian, Phoenician, and Greek stem, and then branches out due to Roman expansion politics (31). As already became apparent in the context of how the *Schöpfungshieroglyphe* guides world history, Herder remains ambiguous about (early) modernity. Still, the tree analogy implies that the *present*'s high and thin branches allow for a better view of the whole (70, 84, 95). It also suggests that new human processes of development may start from the tree's seeds, and that cyclical movements in history co-exist with natural development (42–43; Meinecke 1936, 410). When Ranke speaks of history as a tree, he stresses that the leaves are all interconnected through their common root (Hinrichs 1954, 145; but cf. FHA 4, 56). The tree represents the beauty of diversity; the plurality of nations that are isolated and still ultimately all connected. In *TTPH*, the image of the tree appears in tandem with the *Lebensalter* (life-stages) of humanity (41). This analogy, between how one human being and humanity as a whole develop, shows paradigmatically how individuality and totality are reconciled (16, FHA 9/1, 84). The *Lebensalter* of the "macro-Anthropos" provide a concrete way of conceptualizing humanity's historical development (Heinz 1994, 83). History as described by the analogy is thus progressive—humanity matures, ages, and produces offspring—but the standards for measuring such progress differ with each life-stage.

Relativistic themes

In *Der Historismus und seine Probleme* (*Historicism and Its Problems*), Ernst Troeltsch claims that Ranke managed to avoid pessimism only "by holding on to admittedly pretty obscure leftovers of the Christian faith concerning revelation and salvation" (Troeltsch 1922, 123). In his assessment of the foundational status of Herder's 1774 *TTPH* for German historicism, Frederick Beiser similarly adds the following caveat:

> There is also a shadowy side to Herder's legacy, which bestowed upon the historicist tradition not only central themes but also basic problems. The worst problem was relativism, i.e., how there could be universal standards if all values are cultural and historical. While Herder clearly sees this problem and struggles to avoid it, he offers nothing toward its solution but religious faith, an appeal to providence. It was a desperate strategy; but also the precedent for Humboldt, Ranke and Droysen.
>
> (Beiser 2011, 132)

Beiser and Troeltsch exemplify a general pattern of theorizing about relativism and historicism. According to this pattern, historicists inevitably face the problem of relativism, and every historicist needs to adopt strategies for *countering* this relativism.[11]

This chapter defends two connected claims regarding this supposedly necessary connection between historicism and the problem of relativism. First, Herder and Ranke's theology of history provides the monist basis that underlies their view of history. Thus, Herder and Ranke were no relativists. Second, it is misleading to interpret their theology of history as a "counter-strategy" against relativism. Because of Herder and Ranke's theology of history, the problem of relativism never became pressing in the first place.

As I have emphasized in the above sections, Herder and Ranke were strongly optimistic about the worth of all human history. This optimism was inspired by their theology of history. It may have come to appear as a "desperate strategy" to us. But rather than being "leftovers of the Christian faith concerning revelation and salvation," this theology of history was essential to Herder's and Ranke's conceptions of history. In the remainder of this chapter, I will explain how Herder's and Ranke's theses on *prima facie* relativistic themes—*historicity*, *individuality*, *equal validity*, and *impartiality*—are best understood as *expressions of* religious faith rather than as problems that need to be *countered by* faith. These theses strikingly diverge from what most historical relativists would later claim.[12]

Historicity

While, according to Herder and Ranke, events and societies should indeed be approached as the products of history, still, all historical phenomena are primordially the products of God. And although this dependence upon God's provenience should not lead to strict determinism, and God does not directly interfere in history, it does not leave historical phenomena merely contingent either (cf. FHA 4, 48; Ranke 1828, 198). Rather, concrete historical phenomena express the infinite.

For Herder, the history of human culture is a history of God's education of humanity. Every moral character, no matter how different from ours, expresses a stage in this education and is thus *human* and *moral*. And when the character of a certain people does appear immoral to us, *theodicy* steps in: Herder praises God's *diverse methods* for the education of the human soul and reprimands his contemporaries' shortsightedness (FHA 4, 15, 51). Furthermore, Herder defends practices that were perceived as immoral by his contemporaries in at least three ways, all of which refer to the divine dimension of world history. First, he places practices within their historical context and explains why they appear *less unethical* when appropriately situated (53). This move involves a critique of Herder's contemporaries' tendency to rely exclusively on their own value-system. Herder instead claims that God's wisdom is reflected in the immensely diverse field of human-historical possibilities (106). Second, Herder states that some evils are necessary for historical development or to make room for virtues that would otherwise not have been possible (36, 87–88). Finally, Herder refers to our cognitive limitations with respect to "contingency, fate, divinity": we are ultimately incapable of evaluating God's ways regarding world history in its entirety. This is why frequently we cannot, from our perspective, see beyond the *seeming* contingency of events (58).[13]

More generally, in their theodicies, Herder and Ranke unequivocally adhere to certain permanent core values. The history of humanity has a stable human subject across historical changes; it is this subject that takes on diverse human shapes.[14] Ranke even allows "certain unchangeable eternal main ideas, for instance those of morality"—even though he denies that we can use these ideas to measure moral progress (TP 21–22, AWN 2, 61).

Individuality and incommensurability

According to Herder and Ranke, the historian's task is to trace the development of human practices and to understand them in their own context as well as in comparison with others. As a result of this genetic and comparative procedure, some aspects of these practices appear completely individual and in fact extremely difficult to compare with anything else.[15] One important implication of this undertaking seems to be that cultures are understood as isolated wholes and that their values are centered on their "core." This would make intercultural understanding very difficult at best.

And yet, according to Herder and Ranke's theology of history, God thinks of all individualities as forming an ultimate harmony. The historian, in turn, emulates this ideal in searching for the interconnectedness of universal history, and by attempting to delineate a unified world-historical process. Thus, Herder confidently asserts, against Voltaire's critique of the Middle Ages, that this period was simultaneously "*only like itself,*" an "*individual condition of the world!*" as well as striving "towards a *greater whole*" (PW 207, FHA 4, 51). And Ranke states: "The historian must ... perceive the difference between the individual epochs, in order to observe the inner necessity of the sequence" (TP 22, AWN 2, 62). Herder and Ranke were confident that the sequence of history, just like the course of nature, can ultimately be justified in terms of theodicy. History's providential organization is visible to the same extent as the order of nature is visible: like the book of nature, history is hieroglyphic, but (ultimately) readable. The historian's task is to connect letters and words into a meaningful sequence where "everything occurs in its proper place" (PW 342, FHA 4, 90). The ideal historian would "learn to see the value of ages that we now despise"; "show us a *plan* where we formerly found *confusion*"; and present "prospects of a higher than *human this-worldly existence*" (PW 342, translation modified, FHA 4, 89–90).

Equal value and equal rights

Probably the most notorious aspect of relativism is the idea that all cultural frames are of equal value. This idea can take many forms; here it is useful to distinguish between the metaphysical claim that all value-systems are (in fact) equal, and the methodological imperative to treat them *as if* they were all of equal value.[16] Generally, the idea is considered to be motivated by the relativistic reasoning that since there is no absolute standpoint from which cultures can be ranked, they are all of equal worth. To this reasoning, anti-relativists respond with the "self-refutation argument," which takes the form of the question: If there is no absolute standpoint, then from which perspective could we possibly assert that all cultures are equal?

Ranke would have no problems at all in answering this question. But this is because he does not share one of the relativist's premises: "I picture the deity— since no time lies before it—as surveying all of historic mankind in its totality and finding it everywhere of equal value" (TP 22, AWN 2, 62–63). For Ranke, God's perspective is the standpoint from which equal value is distributed. In modern epistemological terms, this is an absolutist move. In theological terms, it refers to the limitations of our finite condition and to the belief that, unlike us, God is able to appreciate every aspect of creation. Nonetheless, from his theological claim involving an absolute standpoint, Ranke does infer the methodological imperative to treat all of historical humanity in the same manner. For he writes in the same passage: "Before God all generations of men appear endowed with equal rights, and this is how the historian should view matters" (TP 22, AWN 2, 63). These "equal rights" do not demand that the historian actually *sees* all generations as of equal value—for Ranke, this is humanly impossible: human knowers will always have their own standpoint as well. What is more, Ranke himself clearly ranks nations, and even denies some the status of culture (85).

Herder's view is slightly different: He dismisses the idea of a state of nature and claims that every human way of life is cultural. Moreover, he does not adhere to the idea of the *equal* value of cultures (Spencer 2012, 110). However, much like Ranke, he does consider all human cultures to have an autonomous value and to deserve equal rights in historical research. These assertions are made from God's point of view.

Impartiality and empathy

The methodological implications of the notion of "equal rights" deserve separate treatment. Ranke's principle of *impartiality* has come to stand for the idea that historians should "extinguish" themselves so that the objects of historical enquiry can "speak for themselves" (Beiser 2011, 277).[17] Ranke intended to make historiography more objective. Yet, in the debates on historicism and relativism, this ideal of objectivity has been identified with the inability to make any value judgments. Due to the attempt to eclipse all sub-jective elements of judgment, critical judgment becomes impossible. Herder's notion of *Einfühlung* (empathy) has been criticized for the same reason. By fully entering the other's perspective, the historian would become interminably tolerant.[18] Nothing seems left to say except that every party is right according to their own standards.

In reality, Herder and Ranke did consider it possible for historians to evaluate their material. It is useful to distinguish between two forms of moral judgment that are important in interpreting Herder and Ranke. The first form consists of the regular human capacity to criticize injustice. In Herder's case, this happens according to the core values derived from human nature; in Ranke's case, according to the "eternal main ideas of morality." The second form of moral judgment is of a higher sort. Here the historian recognizes the divine nature of human historical existence. These two forms can both be found in Herder's historiographical practice (e.g., FHA 4, 53). Furthermore, Ranke states that the impartial historian approaches any conflict by viewing both sides "on their own ground, in their own environment, so to speak, in their own particular inner state. We must understand them before we judge them" (TP 14, AWN 4, 81). Impartiality thus stands explicitly in the service of evaluation (Beiser 2011, 278–279). Furthermore, complete impartiality has to remain an ideal (Berding 2005, 49). Nevertheless, the historian is meant to emulate a higher ideal of passing judgment. Ranke asks: "We can identify error, but where is there no error? This will not lead us to condemn what exists. Next to the good we recognize evil; but this evil is human too" (AWN 4, 81).

To sum up, Herder and Ranke both identify the following steps in the historian's judgment. First, they recognize that we are all situated and prejudiced. Impartiality and empathy are consequently meant to broaden our horizon so that we expand our understanding beyond the received opinions. Yet, this methodological ideal does not lead *us* to grasping the equal validity of all of historical humanity. We do pass moral judgment after our investigations. Nevertheless, in the end, we have to attempt to recognize the human character and the function in history's providential course even of amoral practices. Thus, ultimately Ranke and Herder both seek to treat worldly evil in a merciful way. This ideal of mercy stems from the belief that God will forgive. For God, the whole of humanity has (equal) value, even if certain periods are obviously less virtuous or pious than others. The historian will never be able to forgive perceived wrongs as mercifully as does God. Still, Herder and Ranke show tolerance toward historical and cultural others precisely because they expect to find absolute value spread out across different forms of human life.

Conclusion

In their philosophies of history, Herder and Ranke share a providential conception of world history and a view of redemption as the uncertain or open telos for humanity. This theology of history informs many aspects of their work. It provides historical individualities with an autonomous significance. And it simultaneously promises that the providential character of universal history can at least be intimated. Finally, it makes the task of the historian akin to that of the priest.

As a result, the starting points from which Herder and Ranke approached themes like historicity, individuality, (equal) validity, and impartiality were different from how philosophers around 1900 assessed historicist theses.[19] Accordingly, I interpret the positions of Herder and Ranke on these themes not as aspects of a relativistic position, nor as strategies for countering a slide into relativism. Herder and Ranke's philosophy of history had a monist foundation in theology, and their theses on the value of history show that they were no relativists. Rather, when Herder and Ranke searched the value of *all* of history, and to judge mercifully, they performed a religious task. This task was to uncover the ultimate meaning of history and to preserve partial hints of the divine for future generations. It was a project that "we" later abandoned. But for Herder and Ranke, it was not a "desperate strategy" against relativism, and it was based on more than mere "obscure leftovers of Christian faith."

Abbreviations

References to Herder are to FHA: *Frankfurter Herder-Ausgabe* (Herder 1985–2000), referring to volumes and pages; translations are from PW: *Philosophical Writings* (Herder 2002). Unless indicated otherwise, references to Ranke are to AWN: *Aus Werk und Nachlass* (Ranke 1964–1975), also referring to volumes and pages; translations are from TP: *The Theory and Practice of History* (Ranke 1973).

Acknowledgments

This work was supported by the European Research Council (Project: The Emergence of Relativism, Grant No: 339382). Many thanks to Martin Kusch and Katherina Kinzel for their comments on various earlier drafts of this chapter.

Notes

1 For other interpretations that highlight the importance of theology in Herder and Ranke, see Pfaff (1983); Menze (2000); Hinrichs (1954); Krieger (1977); and Berding (2005).
2 On Baghramian's list are: empiricism, diversity, respect and tolerance, the identification of reason with practices and language use, tradition, and incommensurability. Relevant aspects on Kusch's list are: dependence, plurality, exclusiveness, (some form of) equal validity, contingency, and tolerance.
3 In this chapter, I will restrict myself to the relevant texts of the early Herder (up until 1774), and I will focus on the methodological reflections in the introductions of Ranke's lecture courses. My comparison of their works is chiefly instrumental to the systematic aim of this chapter. It necessarily glosses over important differences.
4 The genre of "philosophy of history" was still very young, the term being coined in 1765 with Voltaire's "*La Philosophie de l'Histoire*." Speculation about history

in its totality was of course much older, reaching back to the Bible. It was heavily practiced by Enlightenment historians and philosophers alike; Herder considered *This Too a Philosophy of History* a response to a current fad (Koselleck 1975, 658–678; Kondylis 1981, 421–468).

5 These criticisms have been analyzed in great detail by Gjesdal (2017, 151–178) and Sikka (2011, 84–125).

6 References to English translations will be amended with a reference to the corresponding German editions. All other translations are my own.

7 All italics in quotations correspond to the original.

8 Hence, I think that in Herder's case, "genetic explanation" should be differentiated more sharply from the more general notion of "historicization" (see Beiser 1987, 141–42).

9 The difficulties in converging on an interpretation are hence unsurprising: see Bengtsson (2010, 284), Gaier (1988, 73), and Pfaff (1983, 415–417), whose accounts of how the hieroglyph functions as an ordering principle look very different.

10 I modified the translation after consulting Beiser (2011, 281) and Krieger (1977, 13).

11 Cf. Iggers on Herder (1968, 36) and Ranke (78). See also Kinzel's chapter in this volume for how Wilhelm Windelband and Wilhelm Dilthey equated historicism with relativism.

12 I cannot provide the ultimate comprehensive definition of historical relativism, and so I cannot exclude the possibility that connections to some form of historical relativism remain. But the following discussion of various relativistic themes does aim to distinguish Herder and Ranke's theses from a fairly wide net of relativist positions.

13 The providential course of the development does not demand evil as such. It merely demands a general dynamic in history. Hence, even events that seem counterproductive to us must be considered good in some indiscernible way. This doctrine would come to be known as the *Heterogonie der Zwecke* (heterogony of ends) (Meinecke 1936, 414; Kondylis 1981, 631).

14 For a more detailed analysis of how Herder thinks humanity shifts shape throughout history and nonetheless remains the same, see Wildschut (2018).

15 At least this is how I interpret Herder's method. See Wildschut (forthcoming). Ranke seems to posit "individuality" more straightforwardly as a principle of history that opposes the principles of philosophy (AWN 4, 76–77).

16 For a taxonomy of other possible versions, see Kusch (2016, 107).

17 See Kusch's contribution to this volume for Simmel's critique of Ranke on this point.

18 Gjesdal (2017, esp. 154–166) recently presented an extensive critique of this interpretation.

19 See Kondylis (1981, 615–636) for a convincing explanation of how Herder embraced the empirical and historical with *umgekehrten Vorzeichen* because of his radically optimistic ontology.

References

Baghramian, M. (2004), *Relativism*, London and New York: Routledge.

Beiser, F. (1987), *The Fate of Reason: German Philosophy from Kant to Fichte*, Cambridge: Harvard University Press.

—— (2011), *The German Historicist Tradition*, Oxford: Oxford University Press.

Bengtsson, S. (2010), "Challenging Linearity: Finding the Perspective of Herder's Aelteste Urkunde des Menschengeschlechts," in *Herausforderung Herder. Ausgewählte Beiträge zur Konferenz der Internationalen Herder-Gesellschaft*, edited by S. Groß, Heidelberg: Synchron, 281–300.

Berding, H. (2005), "Leopold von Ranke," in *The Discovery of Historicity in German Idealism and Historism*, edited by P. Koslowski, Berlin, Heidelberg, and New York: Springer, 41–58.

Cassirer, E. (1916), *Freiheit und Form: Studien zur Deutschen Geistesgeschichte*, 6th ed., Darmstadt: Wissenschaftliche Buchgesellschaft, 1994.

Crowe, B. D. (2009), "Beyond Theological Rationalism: The Contemporary Relevance of Herder's Psychology of Religion," *Method and Theory in the Study of Religion* 21: 249–273.

Dilthey, W. (1901), "Das Achtzehnte Jahrhundert und die Geschichtliche Welt," in Dilthey, *Gesammelte Schriften*, vol. 3, Göttingen: Vandenhoeck & Ruprecht, 1927, 209–268.

Gadamer, H. G. (1960), *Wahrheit und Methode. Grundzüge einer philosophischen Hermeneutik*, in Gadamer, *Gesammelte Werke 1. Hermeneutik I*, Tübingen: Mohr-Siebeck, 1990.

Gaier, U. (1988), *Herders Sprachphilosophie und Erkenntniskritik*, Stuttgart-Bad Cannstatt: Frommann-Holzboog.

Gjesdal, K. (2017), *Herder's Hermeneutics: History, Poetry, Enlightenment*, Cambridge: Cambridge University Press.

Graczyk, A. (2014), *Die Hieroglyphe im 18. Jahrhundert: Theorien zwischen Aufklärung und Esoterik*, Berlin, München, and Boston: De Gruyter.

Häfner, R. (1995), *Johann Gottfried Herders Kulturentstehungslehre. Studien zu den Quellen und zur Methode seines Geschichtsdenkens*, Hamburg: Felix Meiner Verlag.

Heinz, M. (1994), "Historismus oder Metaphysik? Zu Herders Bückeburger Geschichtsphilosophie," in *Johann Gottfried Herder. Geschichte und Kultur*, edited by M. Bollacher, Würzburg: Königshausen & Neumann, 75–87.

Herder, J.G. (1883), *Aelteste Urkunde des Menschengeschlechts. Erster Band. 1774*, in Herder, *Sämtliche Werke*, vol. 6, reprint, Hildesheim: Olms, 1967, 193–511.

—— (1985–2000), *Werke in zehn Bände*, edited by G. Arnold et al., Frankfurt am Main: Deutscher Klassiker Verlag.

—— (1993), *Against Pure Reason: Writings on Religion, Language, and History*, translated and edited by M. Bunge, Eugene, OR: Wipf and Stock Publishers.

—— (2002), *Philosophical Writings*, translated and edited by M. N. Forster, Cambridge: Cambridge University Press.

Hinrichs, C. (1954), *Ranke und die Geschichtstheologie der Goethezeit*, Göttingen: Musterschmidt.

Iggers, G. (1968), *The German Conception of History. The National Tradition of Historical Thought from Herder to the Present*, 2nd ed., Middletown, CT: Wesleyan University Press, 1983.

Kondylis, P. (1981), *Die Aufklärung im Rahmen des neuzeitlichen Rationalismus*, reprint, Hamburg: Meiner Verlag, 2002.

Koselleck, R. (1975), "Die Herausbildung des modernen Geschichtsbegriffs," in *Geschichtliche Grundbegriffe: Historisches Lexikon zur politisch-sozialen Sprache in*

Deutschland, vol. 2, edited by O. Brunner and R. Koselleck, Stuttgart: Klett-Cotta, 647–717.

Krieger, L. (1977), Ranke *the Meaning of History*, Chicago: University of Chicago Press.

Kusch, M. (2016), "Relativism in Feyerabend's later writings," *Studies in History and Philosophy of Science* 57: 106–113.

Löwith, K. (1949), *Meaning in History*, Chicago and London: University of Chicago Press.

Meinecke, F. (1936), *Die Entstehung des Historismus*, 2nd ed., München: Leibniz Verlag, 1946.

—— (1942), *Aphorismen und Skizzen zur Geschichte*, Leipzig: Koehler & Amelang.

Menze, E. A. (2000), "'Gang Gottes über die Nationen': The Religious Roots of Herder's *Auch eine Philosophie* Revisited," *Monatshefte* 92: 10–19.

Pfaff, P. (1983), "Hieroglyphische Historie. Zu Herders *Auch eine Philosophie der Geschichte zur Bildung der Menschheit*," *Euphorion* 78: 407–418.

Ranke, L. (1949), *Das Briefwerk*, edited by W. P. Fuchs, Hamburg: Hoffmann und Campe Verlag.

—— (1964–1975), *Aus Werk und Nachlass*, edited by W. P. Fuchs and T. Schieder, 4 vols., München: R. Oldenbourg Verlag.

—— (1973), *The Theory and Practice of History*, 2nd ed., edited by G. Iggers, translated by W. Iggers, London: Routledge, 2011.

Sikka, S. (2011), *Herder on Humanity and Cultural Difference: Enlightened Relativism*, Cambridge: Cambridge University Press.

Spencer, V. A. (2012), *Herder's Political Thought: A Study of Language, Culture, and Community*, Toronto: University of Toronto Press.

Troeltsch, E. (1922), *Der Historismus und seine Probleme*, in Troeltsch, *Gesammelte Schriften*, vol. 3, Tübingen: Mohr-Siebeck, 1977.

Wildschut, N. (forthcoming), "Analogy, Empathy, Incommensurability. Herder's Conception of Historical Understanding," in *Herder on Sympathy and Empathy: Theory and Practice*, edited by L. Lukas, E. Piirimäe, and J. Schmidt, Leiden: Brill.

—— (2018), "Proteus and the Pyrrhonists. Historical Change and Continuity in Herder's Early Philosophy of History," *Herder Jahrbuch / Herder Yearbook* 14: 65–93.

2 The history of philosophy and the puzzles of life

Windelband and Dilthey on the ahistorical core of philosophical thinking

Katherina Kinzel

Introduction

Toward the end of the nineteenth century, the German philosophical community struggled with a philosophical "identity crisis" (Schnädelbach 1984, 5–11). The success and proliferation of the special sciences had raised fundamental questions concerning the objects, methods, and social functions of philosophy vis-à-vis empirical science. Additional pressure came from the professionalization of the study of history. Since the early nineteenth century, historians had become increasingly vocal about their authority over the interpretation of all things historical. The human-historical world, they insisted, should be studied by empirical historiography, not by philosophy, which tended to impose distorting schemes on the historical process (see, e.g., Ranke 1831–1832, 74–78).

Philosophers, in turn, drew connections between "historicism" and "relativism." They worried that, when applied to philosophical questions, the methods of professional history would turn into a destructive force undermining the basis for rational and normative discourse (Windelband 1883; Husserl 1911; Rickert 1924). Taking up a slogan that was first formulated by Ernst Troeltsch and Karl Heussi in the 1920s and 1930s (Troeltsch 1922b; Heussi 1932), Charles Bambach describes the period after 1880 as marked by a "crisis of historicism."[1] The rise of historical thinking had philosophers grapple with "problems of epistemological nihilism, cultural dissolution, and historical relativism" (Bambach 1995, 53).

When speaking of historicism, philosophers usually referred to the historicization of beliefs and values in general. Less attention was devoted to the historicization of philosophical systems and doctrines. Nevertheless, a "crisis of the historicization of philosophy" occurred within the broader "crisis of historicism." As philosophers began to emulate the methodological rigor of professional historians when studying the past of their field, they developed an acute sense for the cultural determinants of philosophical thinking (Geldsetzer 1968; Hartung 2015). The simultaneous decline of speculative metaphysics diminished the *prima facie* plausibility of progressivist narratives about the history of philosophy. The thought emerged that all philosophical systems and their supposedly universal knowledge claims might be relative to culture, national context, and historical period.

This chapter analyses the contribution of the historicization of philosophy to late nineteenth-century debates over relativism. It studies the "crisis of the historicization of philosophy," and the dominant philosophical strategies against it, by focusing on two key figures of the period—Wilhelm Windelband and Wilhelm Dilthey.

Windelband's and Dilthey's respective philosophical projects can be read as two paradigmatic attempts of dealing with the problems and pressures that, at the time, were shaping philosophical discourse. Both were concerned with clarifying the relations between the various special sciences, and saw a role for philosophy in illuminating their methodological foundations. Both endorsed non-metaphysical philosophies and evoked Kantian criticism as a model for how philosophical analysis should proceed. Both believed in the significance of history for revealing the meaning of human life and culture and sought to revise the Kantian approach such as to make room for the acknowledgement of humanity's essential historicity. And both struggled with the question as to how historical insights can be reconciled with the supposedly universal knowledge claims of philosophy.

Analyzing the similarities and differences between Windelband and Dilthey, this chapter shows that both thinkers' meta-philosophical reflections were shaped by a concern with historical relativism. It argues that despite the fundamental differences between their philosophical projects, Windelband and Dilthey both answered to a perceived "crisis of the historicization of philosophy." It uncovers the striking similarities in their proposed answers: by defining philosophy in terms of the relationship between conceptual thinking and life, both authors sought to uncover an ahistorical core of philosophical thinking, and in this way, to block the route to relativism. The chapter concludes that, although Windelband and Dilthey succeeded in averting historical versions of relativism, they did so at the cost of incurring a relativism *vis-à-vis* philosophical systems. This relativism turned out to be rooted not in the historicity of philosophy but in the timeless essence of philosophical reasoning itself.

Transcendental absolutism and immanent universalism

Windelband and Dilthey not only clashed over the demarcation of the natural and the human-historical sciences (Dilthey 1883, 1895–96; Windelband 1894). More fundamentally, their views on the goals and methods of philosophy were radically different if not diametrically opposed. It would be premature to describe this conflict in terms of the opposition between Kantianism and historicism. While Windelband indeed positions his transcendental philosophy against the full-on historicization of reason that, in his view, leads to relativism, he also tries to incorporate central insights of historicism into his philosophy (Beiser 2008; Kinzel 2017; Ziche 2015). Conversely, Dilthey's "critique of historical reason" (Dilthey 1883, 165) combines an acknowledgment of the inherent historicity of reason with the recognition that historical knowledge needs a strong epistemological grounding so as not to disintegrate into relativism (Makkreel 1975, 53). The historicist elements in Dilthey's philosophy are balanced by a strong universalism about the structure of human life and experience. In the following, I seek to capture the fundamental differences between Windelband's and Dilthey's philosophical projects not in terms of Kantianism and historicism but in terms of what I will call *transcendental absolutism* and *immanent universalism*. As I will show, this opposition implies further differences in how Windelband and Dilthey address the problem of relativism.

Windelband's brand of Neo-Kantianism answers to the perceived identity crisis of philosophy by saving it a special role "above" the sciences (Beiser 2014, 492–500). As a "critical science of universal values" (Windelband 1882b, 29) philosophy is a second order discipline. It determines the "system of axioms" (Windelband 1883, 107)—the system of universal and necessary values—that governs discourse in science, culture, and politics, thus securing the possibility of rational discourse in these areas.[2]

Central to Windelband's philosophical project is a distinction between the factual and the normative. He introduces this distinction as a formal one between two fundamental and irreducible types of cognitive operations: judgments and evaluations. While judgments involve a synthesis of ideas, evaluations express the relation between the evaluating consciousness and the represented object (Windelband 1882b, 29). But for evaluations to have normative force, they need to express not just the subjective attitude of the evaluating subject. Rather, according to Windelband, the validity and normative force of particular evaluations derive from their correspondence to absolute values. Accordingly, there is a system of *absolute*—that is, necessary and universal—values, and this system is grounded *transcendentally*. Windelband also refers to this system as "normal consciousness," which is distinct from and determines empirical consciousness (Windelband 1882a).[3]

We may refer to this position as *transcendental absolutism*. Transcendental absolutism sets philosophy the goal of revealing the system of absolute norms. Windelband stresses that philosophy can do so by purely formal means, because absolute norms are teleologically necessary. They are "norms which ought to be valid if thinking wants to fulfill the purpose of being true, volition the purpose of being good, and feeling the purpose of capturing beauty, in a manner that warrants universal validation" (Windelband 1883, 109; see also Windelband 1882a, 74). It is only when drawing on a formal, non-empirical method that philosophy can preserve the essential distinction between the factual and the normative.

It is with respect to both the issue of transcendentalism and the distinction between the factual and the normative that deep differences arise between Windelband and his contemporary Dilthey. As many commentators have stressed, Dilthey explicitly sought to overcome the dualisms of

Kantian philosophy. He questioned the distinction between intuition and understanding, between theoretical and practical philosophy, as well as that between the transcendental and the empirical (Ermarth 1975, 149–152; Makkreel 1975, 223–224). For Dilthey, philosophy is not a formal enterprise but a more hybrid endeavor. While philosophy does play a foundational role for the "human sciences," its methods are continuous with those of empirical disciplines, in particular with those of descriptive psychology, anthropology, and history (Patton 2015).[4]

Hence, Dilthey's philosophical project is "transcendental" only in a loose sense. While his foundational enterprise of the 1880s and 1890s concerns itself with the conditions of possibility of knowledge in the human sciences, Dilthey finds these conditions not in a transcendental realm but rather contained in the dynamic structure of "lived experience." Dilthey's central concept in this context is the "psychic nexus," which integrates the human faculties, "the intellect, the life of the drives and feelings, and the activity of the will" (Dilthey 1894, 154) into a structured totality. This totality is relational because the "life-unit" finds itself conditioned by a surrounding "milieu" and conversely acts upon this "milieu" (Dilthey 1894, 172). At the same time, it is also purposive and teleological, because all processes of psychic life are directed toward reaching a state of congruence with the surrounding milieu (Dilthey 1894, 183).

Based on this conception of a dynamic, purposive, and self-adapting "psychic nexus" Dilthey objects to Kant's distinction between epistemology and psychology, as embraced by Windelband.[5] He argues that the synthesis of the facts of consciousness cannot be achieved by a transcendental method but only by the living totality of the psychic nexus (Dilthey 1894, 126). Therefore, the principles of knowledge should be sought by way of "self-reflection" (Dilthey 1880–1890, 278). Different from traditional epistemology, "self-reflection" attends to the totality of psychic life (see also Lessing 2016, 105–106). It grasps lived experience in its raw state of primary connectedness and seeks to disclose the principles of reason *from within* this dynamic totality. Dilthey's philosophy is, at its core, a philosophy of *immanence*.[6]

Consequently, normativity, too, is rooted in the living psychic nexus. Dilthey argues that all values are first and foremost "life-values." They emerge when the "life-unit" represents the enhancing or inhibiting effects of the surrounding milieu by attaching to it positive and negative feelings (Dilthey 1894, 177). Dilthey also allows for values to develop historically.[7] The dynamic activity of the psychic nexus creates new values, and the moral-historical development of humankind leads to the development of "higher" values (Dilthey 1894, 180–181).

Nevertheless, the historicist elements in Dilthey's thinking are balanced by strong universalist convictions. Dilthey postulates a human commonality that "is expressed in the selfsameness of reason, in sympathy as part of the life of feeling, and in the mutual commitments of duty and justice" (Dilthey 1910, 163). He also argues that the differences between individuals are explicable in terms of mere quantitative differences in their character traits (Dilthey 1894, 151–152). It is unclear whether he believes in the universality of values, given that he thinks of them as historical entities (see also Makkreel 1975, 243). But he clearly holds that all values, even those that vary from culture to culture, are rooted in a common ground—they can all be traced to the dynamic structure of the living psychic nexus. This structure is indeed universal. We may refer to this position as *immanent universalism.* While ethical values and the principles of reason are strictly immanent to lived experience, they also share a universal basis in the dynamic structure of psychic life.

The challenge of relativism

Unsurprisingly, the deep disagreements between Windelband's and Dilthey's philosophical projects also translate into differences in their approaches to relativism. Both authors draw connections between certain forms of historical thinking and relativism. But they differ on how exactly historical thinking causes relativism and, consequently, on what a solution to the problem would look like.

As emphasized above, Windelband insists on a firm distinction between the factual and the normative. And in his view, relativism emerges precisely when this boundary is violated: it results from the attempt to derive the "system of values" not from formal considerations but from empirical facts about human psychology or history. Windelband gives two reasons for why this strategy is futile. First, empirical generalization from psychological or historical facts are only approximations. They can never reach the universality and necessity that is characteristic of absolute norms (Windelband 1883, 114). Second, empirical generalization cannot discriminate between true and false, or right and wrong. Psychological laws cause both true and false beliefs, and historical cultures embrace both right and wrong behaviors. Hence, for the genetic method, all beliefs and actions have to appear as equally justified. "For [the genetic explanation], there is no absolute measure; it must treat all beliefs as equally justified because they are all equally necessary by nature" (Windelband 1883, 115). Windelband concludes that "*relativism* is the *necessary consequence*" of the purely empiricist treatment of philosophy's cardinal question" (Windelband 1883, 116). Note though that is not historical thinking as such that causes relativism but rather historical thinking that oversteps its boundaries to address non-empirical issues of justification and normativity (see also Kinzel 2017).

From Windelband's perspective, Dilthey's philosophy of immanence seems like a clear case of the genetic method overstepping its boundaries. Indeed, Windelband accuses Dilthey of relativism, albeit indirectly. He urges that the "critique of historical reason" needs to draw on absolute values: "it has to be Criticism, and as such it needs a standard" (Windelband 1883, 122). The implication is that Dilthey's philosophy lacks such a standard and that for this reason, it leads to relativism.[8]

And yet, Dilthey does not see himself as a relativist. He thinks of his foundational project as preserving and explicating the validity of knowledge in the human sciences (Dilthey 1883, 50), he presents his *Weltanschauunglehre* (theory of world-views) as a solution to the problem of relativism (Dilthey 1907, 66), and in his famous *Rede zum 70. Geburtstag* (*70th Birthday Address*) of 1903, he declares his lifelong devotion to solving the conflict between "historical consciousness" and universal knowledge claims (Dilthey 1903, 9).[9] For Dilthey, it is not the violation of the boundary between the factual and the normative that leads to relativism. Instead, relativism emerges as a by-product of the historicist destruction of human universals.

As Dilthey observes, the historical perspective reveals a never-ending conflict of belief systems, with no prospect of eventual resolution. In his view, the mere empirical fact of unresolved conflict already calls into question the "objective validity" of each particular world-view (Dilthey 1960a, 3). But more than that, insight into the variability of human forms of life also calls into question the existence of universal human commonalities. As Dilthey puts it, "the human type came apart in this process of development" (Dilthey 1960b, 77, see also Dilthey 1960a, 6). Taken together, these two factors—undecidable conflict and lack of human commonalities—lead to relativism. The historical-developmental perspective

> is necessarily connected with knowledge about the relativity of every historical form of life. The absolute validity of any particular form of life, constitution, religion or philosophy disappears before the view that embraces the whole world and all pasts.
>
> (Dilthey 1960b, 77)

Note also that for Dilthey, it is not history as a method or a form of explanation that causes relativist problems. Rather, is it by providing a synoptic grasp of all pasts—a survey of the *totality* of human differences—that history undermines the belief in human universals and hence leads to relativism. Because Dilthey gives a different account of the causes of relativism than Windelband, he can also propose a different solution. In his view, a stronghold against relativism needs to be built not from transcendentalism but from universalism. What is required is not a formal method for uncovering absolute values but rather an account of the universal structure of human life.

Philosophy and the puzzles of life

Despite the deep differences between their philosophical projects, and despite the fact that, in line with these differences, Windelband and Dilthey promise different solutions to relativism, in their meta-philosophical reflections, a surprising convergence occurs. Here, the two authors respond not only to the historicization of life and culture in general. Rather, they address the more specific questions that are raised by the historicization of philosophical thinking.

By the mid nineteenth century, the professionalization of the study of history had made possible a new way of thinking about the history of philosophy: the thought had emerged that philosophy itself might be relative to time, historical culture, and nationality (see Ranke 1831–1832, 76). The simultaneous demise of speculative metaphysics had scattered philosophers' confidence that the historical variance of philosophical systems could be viewed in terms of the teleological self-realization of reason. Emulating the methodological rigor of professional historians, philosophers' accounts of the history of their field had become less biased, philologically more accurate and more context-sensitive (Hartung 2015). As a result, philosophers themselves were more prone to viewing the history of philosophy in the same way historians did: as contingent and relative.

Being practicing historians of philosophy themselves, Windelband and Dilthey were acutely aware of a "crisis of the historicization of philosophy" that occurred within the broader "crisis of historicism." In their meta-philosophical reflections, both authors address the fact that the methods, subject matters, and goals of philosophy have undergone massive changes over the course of history. Both raise the question as to whether there even is a general conception of philosophy that is applicable throughout all history (Dilthey 1907, 24–25; Windelband 1882b, 6–19; 1891, 3–5).[10] And both seek out an affirmative answer to this question by pointing to an ahistorical core of philosophical reasoning. As I will show, this answer is also meant to solve the conflict between the historicity of philosophical systems on the one hand, and philosophical claims to universal validity on the other. In this way, it is meant to block the road to historical relativism about philosophy.

Dilthey's account of the ahistorical "essence" of philosophy is based in his philosophy of life, which, in turn, takes up and elaborates some of the psychological concepts that he developed in earlier years. In particular, the concept of "lived experience" remains central in Dilthey's meta-philosophical account. Here, too, Dilthey stresses that the dynamic system of psychic life is teleologically driven toward adapting the "life-unit" and the surrounding "milieu" (Dilthey 1907, 34–35; 1960a, 15–16). In this dynamical process, beliefs and life values are constantly put to the test and adapted to changing circumstances. But Dilthey thinks that life needs stability, which can only be gained by transforming the changing world of apperceptions and evaluations into valid, justified knowledge (Dilthey 1907, 35). "Life-experience," then, is the consolidated and stable interpretation of experience that emerges when an individual's drives, feelings, and the objects that they are attached to become encapsulated in general knowledge (Dilthey 1960b, 79). Life-experience also achieves a supra-individual status. It is extended in custom and gains in clarity and certainty as it is handed down throughout the ages in tradition (Dilthey 1960b, 79–80).

And yet, the different elements that compose life-experience are not always coherent with one-another. "The soul tries to compose into a whole the life-relations and the experiences that are based in them, and fails ... Strange contradictions arise, which become ever more apparent in life-experience and which are never resolved" (Dilthey 1960b, 80–81). The stable interpretation that is given in and by the knowledge of "life-experience" is always threatened by inherent tensions. Dilthey refers to these tensions as "world- and life-puzzles" (Dilthey 1907, 346).

World-views then, are attempts to solve these life-puzzles, and philosophy is—among religion and poetry—one of the three domains in which such world-views are formulated. Philosophy differs from the other domains in that it is concerned with justification and generalization. Dilthey thus defines philosophy primarily by way of its functions, the function of testing taken for granted presuppositions, of justifying their universal validity, and of connecting the different elements of life in a coherent system (Dilthey 1907, 61–62). Dilthey has now reached an answer to the question as to what constitutes the essence of philosophy. This essence consists in the application of rigorous conceptual thinking to the puzzles of life.

The notion of philosophy that is present in Windelband's *Einleitung in die Philosophie* (*Introduction to Philosophy*) (1914) is surprisingly close to Dilthey's. While Windelband is far from developing a philosophy of life, or a typology of world-views, and puts a much stronger emphasis on the rational dimension of philosophical thinking than Dilthey, both authors attribute a similar function to philosophy.

The starting point of Windelband's *Einleitung* is the observation that everyday life and culture already contain general views about the world and the concepts to express these views. There is a "metaphysics of the nursery and fairy tale," "world-view of religious dogma," and the "life-picture that we enjoy in the work of the poet and the artist" (Windelband 1914, 2). In addition, the special sciences take up ideas that have been formed in everyday life and transform them into scientific concepts. The subject matter of philosophy initially consists in the "thought-content which is provided by life itself and by the insights of the special sciences" (Windelband 1914, 6).

But philosophy does not just build on life-experience and its concepts. Instead, it critically assesses these ideas and concepts. At the beginning of philosophical reflection stands an experience of shock or unsettling in which taken for granted assumptions collapse (Windelband 1914, 8). This experience calls for a critical assessment of everyday ideas and concepts. According to Windelband, the essential function of philosophy thus consists in questioning and rethinking the concepts and ideas of everyday life and science. By thinking through these concepts, philosophy tries to re-achieve stability and certainty. It seeks to provide a theoretical and practical footing in the world (Windelband 1914, 19).

While Dilthey's central concept is that of the world-view, Windelband thinks of philosophy in terms of its problems. According to Windelband, philosophical problems spring "with objective necessity" from the "vigorous and uncompromising rethinking of the preconditions of our spiritual life"

(Windelband 1914, 8). While he does not give a very detailed account of how this process of rethinking creates "philosophical problems," there are some suggestive hints. He writes that the conceptual presuppositions that philosophy critically engages are often unproblematic in the domains in which they have first emerged. More specifically, they are unproblematic as long as they remain isolated from one another. But like Dilthey, Windelband stresses that philosophy strives for connection, unity and justification (Windelband 1907, 11). The emergence of the philosophical problem is tied to the demand for organizing a complex network of ideas and concepts in a coherent and unified manner. And the "inadequacy and contradictory imbalance" of the contents given to philosophy by life and science creates the "age-old puzzles of existence" (1891, 10). The parallels to Dilthey's "life-puzzles" that arise from the inherent contradictions of life-experience are evident. For both authors, the concepts that emerge from life are contradictory and lead to existential puzzles. They need to be integrated into coherent systems by rigorous philosophical effort.

The ahistorical core of philosophical thinking

As we have seen, Windelband and Dilthey share a definition of philosophy as consisting in the application of rigorous conceptual thinking to the puzzles of life. But the most important agreement between them is that philosophy has an ahistorical core. For both authors, there are some basic structures—world-views in the case of Dilthey, philosophical problems in the case of Windelband—that reoccur throughout the history of philosophy. And therefore, there can be no real progress in philosophy (Windelband 1891, 7, 10; 1914, 10, 17; Dilthey 1907, 65). This thought, the idea of an ahistorical core of philosophical thinking, preserves a general notion of philosophy that remains constant through historical change.

Windelband argues that by rethinking the concepts of everyday life and science, philosophy also expresses the "necessary forms" of reason (Windelband 1891, 8). He refers to Hegel to elucidate the point. Hegel was right, he argues, that the categories of reason express themselves in the historical process. But he was wrong to assume that the succession of philosophical systems follows a necessary trajectory that carries systematic significance (Windelband 1891, 9, see also 1905, 176–177). Windelband thinks that the history of philosophy is shaped by a combination of rational and cultural factors—"temporal causes and timeless reasons" (Windelband 1905, 189).

As recapitulated above, Windelband's *transcendental absolutism* posits an absolute, and hence timeless, system of values that can be discovered by philosophy. But in his historiographical texts, Windelband identifies the ahistorical stratum of philosophical reasoning with problems, not with absolute values: despite historical change, the same "age-old puzzles of existence" reoccur again and again. "Certain differences of world- and life-attitudes reoccur over and over again, combat each other and destroy each other in mutual dialectics" (Windelband 1914, 10). Windelband argues that this persistence of philosophical problems is an indication of them being given "necessarily" and "objectively" (Windelband 1914, 11).

However, he also suggests that philosophical problems—and, one needs to add, problem-solutions—are necessary because the "material" of ideas and concepts given by life and the special sciences contains "the objective presuppositions and logical coercions for all rational deliberation about it" (Windelband 1891, 10). The necessity of philosophical problems is thus a form of logical necessity that derives from the normative laws of rational thinking. Hence, the "system of absolute norms" is part of the explanation as to why philosophical reasoning has an ahistorical stratum. Because philosophical problems are a product of the application of the "timeless" determinations of reason, they themselves have an ahistorical character: the "stock of philosophical problems" is the "necessary content of rational consciousness in general" (Windelband 1914, 17). The goal of the historiography of philosophy, and its contribution to systematic philosophy, consists in separating the historically contingent from the ahistorical, essential, and necessary (Windelband 1905, 189).

The common depiction of Dilthey as an unfailing historicist obscures the fact that he, too, saw the goal of the historiography of philosophy in revealing an ahistorical essence of philosophy. Just like Windelband, Dilthey conceptualizes the history of philosophy as a combination of contingent and timeless factors. But in line with his *immanent universalism* Dilthey has the historiography of philosophy search not for absolute norms but rather for the universal ground on the basis of which conflict and historical change occur (Dilthey 1960a, 13).

According to Dilthey, there are three basic world-view-types that permeate all the different and historically specific formations of philosophical thinking (Dilthey 1960b, 86). Dilthey gives the following rationale for restricting his typology to three basic types: philosophical thinking seeks to grasp the world only under one of the three basic categories of life: knowledge, feeling, or volition. Because of its conceptual nature, it tries to derive a complete system from one of these categories, but this system cannot fully represent all aspects of life in their interconnectedness (Dilthey 1907, 65–66). This gives rise to three basic world-view-types. "Naturalism" tries to make life intelligible from the standpoint of causal knowledge, "objective idealism" is dominated by feeling and understands the world from the perspective of value and meaning, and "idealism of freedom" makes the autonomy of the self and volition into the central categories of its world-view (Dilthey 1960b, 100–118).

This argument remains somewhat sketchy, and Dilthey admits that the typology is tentative and based on intuition as well as on his experience as a historian of philosophy (Dilthey 1960b, 99). But whether by way of explicit argument or intuition, the goal of the typology is straightforward. It serves to show that even when philosophical views are shaped by contingent historical factors, there is nevertheless an essential continuity that unites all philosophical systems. By expressing one of three basic types, each philosophical system participates in the ahistorical nature of philosophical reasoning.[11]

There are important differences between Windelband and Dilthey. Windelband emphasizes the rationality of philosophical thinking, Dilthey emphasizes its limitations *vis-à-vis* the complexity of life. Windelband puts the emphasis on normativity, while Dilthey puts a stronger focus on the dynamic and productive capacities of life. Windelband's account is consistent with a *transcendental absolutism* about norms, while Dilthey's fits with his *immanent universalism*.

And yet, both accounts serve the same goal. They seek to reveal a core of philosophical reasoning that is immune to historical change. This yields not only a general concept of philosophy that is applicable to all historical periods, it also promises a solution to the conflict between philosophy's historicity and its claims to universal validity. Dilthey addresses this promise explicitly. He declares that the relativity of world-views is not the "last word" of the historical mind that has studied them all and that his own theory of world-views presents "in opposition to relativism, the relation of the human mind to the riddle of the world and of life" (Dilthey 1907, 66). For Windelband, too, the historiography of philosophy can avert the "hopeless relativism," which means "the end of philosophy" (Windelband 1905, 187). It can do so if it separates temporal causes and timeless reason, contingent cultural factors and the "normative determinations" that shape philosophical thinking at all times (Windelband 1905, 189). The shared assumption is obviously that the road to relativism can be blocked if philosophy is revealed to have an ahistorical core.

An ahistorical relativism

In very general terms, both Dilthey and Windelband can claim to have avoided relativism about the history of philosophy. To the extent that Dilthey has shown that there is a universal structure of life that is the basis for all conflicting philosophical systems, he has blocked what on his own account is one of the main motivations for relativism. He has blocked the dissolution of universal human commonalities. And to the extent that Windelband has shown philosophical problems to emerge from the timeless determinations of reason, he has upheld his conception of an absolute, non-relative system of values.

Something about this response seems unconvincing though. Rudolf Makkreel finds the theory of world-views to be "the least satisfactory part of Dilthey's philosophy" (Makkreel 1975, 345). In his view, the analysis of world-views is based on psychological concepts that are more static than those employed in his other writings (349). He thinks that Dilthey is incapable of solving the problem of relativism, because his world-views express "fixed cognitive differences" and resist all transformation (352). Michael Ermarth offers a more positive assessment of the theory of world-views. He thinks of Dilthey's world-views as coherent and stabilizing but not self-enclosed or static. They are driven by an inner dialectic that compels them to revise their initial premises (Ermarth 1975, 329–332). Ermarth also emphasizes the sharp distinction between world-views and hermeneutical science. While world-views are relative, the positive knowledge about them is not (334).

Ermarth is right to emphasize the distinction between the level of world-views and the level of their analysis. While this raises some questions of reflexivity whether Dilthey's can claim his own project to classify as philosophy without relativizing it, it allows us to appreciate that what Dilthey's critics took to be breakdown of reason, he himself saw as a breakthrough of historical consciousness (Ermarth 1975, 324). Dilthey indeed believes that by showing how the plurality of philosophical systems is rooted in life, the historical consciousness can resolve "the harsh contradiction between the claim to universal validity in every philosophical system and the historical anarchy of these systems" (Dilthey 1960b, 78). By explaining how all the different philosophical systems and the conflicts between them come about, the theory of world-views occupies a standpoint above them and thus reaffirms the sovereignty of reason (Dilthey 1907, 41, 66). Relativism is avoided for the universal historical standpoint. Johannes Steizinger has emphasized that in making this move, Dilthey also ends up attributing to the theory of world-views qualities that he denies to metaphysical philosophy: it is objective, all-encompassing, and expressive of the sovereignty of reason (Steizinger 2017, 240–242).

However (and this is central), the problem of relativism has not been avoided on the level of philosophical systems. Ultimately, Dilthey's theory of world-views defines philosophy in a way that attributes only relative validity to its claims. Each philosophical system justifies a world-view, expressing the manifoldness of life in a one-sided conceptual system that has only relative validity. The conflict between different, only relatively valid, world-views is eternal, just as the different world-view types are. By making philosophy into something ahistorical, Dilthey has thus merely removed the historical element of relativism. In his analysis, relativism is not a result of the historicization of philosophy but rather built into the ahistorical relation of conceptual thinking to the puzzles of life.

It is no accident that Windelband's account of philosophical problems has the same effect. As reconstructed above, Windelband thinks that philosophical problems are the necessary product of the norms of reasoning being applied to the conceptual materials found in life and science. For this reason, they reoccur throughout history. But while the system of absolute values that comprises the norms of rational thinking is absolute, the conflicting problem-solutions are not. Windelband does not envision his system of absolute values to lead to unequivocal decisions between different philosophical systems. Quite to the contrary, he, like Dilthey, accepts a picture of endless conflict.

Lutz Geldsetzer suggests that in branding genuine philosophical problems as unsolvable, philosophical problem-history is the "expression of modern relativism and scepticism" (Geldsetzer 1968, 167). Indeed, Windelband has done nothing to avoid the conclusion that the different philosophical systems that attempt opposing problem-solutions have only relative but not absolute validity. Like Dilthey, he has made reoccurring conflict between philosophical systems that cannot claim absolute validity into a feature of the timeless structure of philosophical discourse.

Let us take stock. Both Dilthey and Windelband seek to reveal an ahistorical core of philosophical thinking. Both introduce something that grounds, and to some extent explains, the historical variety of conflicting philosophical systems—the universal structure of life in the case of Dilthey and an absolute system of norms in the case of Windelband. Both assume that the introduction of a universal or absolute *explanans* for philosophical conflict solves the problem of historical relativism. However, they both seem to miss that the *explanandum* in question—the conflicting philosophical systems— is not thereby rendered universal or absolute. The conflicting philosophical systems that emerge from the universal structure of life or from the absolute norms of reason have only relative validity. While the historical character of the *explanans* has been removed, relativism about the *explanandum* has not. Ironically, by de-historicizing philosophy, Windelband and Dilthey make the conflict between philosophical systems, each of which has only relative validity, into an eternal feature of philosophical discourse.

This finding not only reveals that Windelband and Dilthey remained somewhat inconsequential in their dealings with relativism, it also attests to how firmly they and possibly many other late nineteenth- and early twentieth-century philosophers—Rickert, the early Husserl and Troeltsch come to mind—associated historicism and relativism with one another. To Windelband and Dilthey, it seemed that the solution to the problem of relativism had to run via a solution to historicism. Conversely, the common expectation was that a solution to historicism would also block the road to relativism. Hence, it was assumed that the "crisis of the historicization of philosophy" could be solved by recourse to an ahistorical core of philosophical thinking. And because they equated historicism and relativism with one another, Windelband and Dilthey did not realize that, when immunizing philosophy from historicism, they incurred a relativism about philosophical systems: a relativism rooted not in the historicity of philosophy but in the timeless essence of philosophical thinking itself.

Acknowledgments

For insightful comments on an earlier draft of this chapter, I want to thank Martin Kusch, Johannes Steizinger, and Niels Wildschut. This work was supported by the European Research Council [Project: The Emergence of Relativism, Grant number: 339382].

Notes

1   Note that while Troeltsch and Heussi declared historicism itself to be in crisis, later commentators, including Bambach, suggest that it was in fact academic philosophy that felt troubled by the implications of the thorough historicization of human

life, culture, and reason (see also Beiser 2015, 24). Similar diagnoses have been put forward by Annette Wittkau (1992) and Otto Oexle (2007). However, Wittkau and Oexle focus not primarily on how the "historicism" debates unfolded in academic philosophy, but reconstruct the broader intellectual and cultural history.

2 Windelband uses the terms "axioms," "norms," and "values" almost interchangeably, since it is not just the logical presuppositions of correct thinking that he has in mind, but also moral norms and rules of aesthetic judgment.

3 Even though Windelband seeks to avoid attributing metaphysical status to the realm of values, he does not go as far as Rickert to claim values to be "unreal" (Rickert 1921). The question of whether the realm of the normative has to be understood as a transcendental reality remains unresolved in Windelband.

4 Dilthey's complex and multifaceted *oeuvre* cannot be measured with one yardstick. There is serious debate concerning the unity and continuity of Dilthey's writings. I cannot go into these issues in detail here, but I agree with Frithjof Rodi that Dilthey's work is united by a set of common themes and argumentative patterns (Rodi 2003). I focus my paper on Dilthey's conception of lived experience and the psychic nexus because of the continuity in these concepts: they are present already in the psychological writings from the 1880s and 1890s and reappear in the philosophy of world-views from the 1900s.

5 This has earned Dilthey the charge of "psychologism." Heinrich Rickert rejected Dilthey's conception of *Geisteswissenschaften* because it did not properly disentangle the epistemology of history from individual psychology (Rickert 1929, 122–127). Troeltsch, too, criticized Dilthey's "psychologism," which in his view consisted in the attempt to derive knowledge about the spiritual-historical world from an "empiricism of immediate experience" (Troeltsch 1922a, 782, 791–793).

6 Rodi puts a strong emphasis on the thought of immanence in Dilthey's philosophy and on this basis also highlights some of the continuities that unite Dilthey's thinking from the 1880s to the 1900s (Rodi 2003).

7 Jacob Owensby has given a detailed reconstruction of Dilthey's views on the relation between the individual life-unit and the socio-historical world and provides a provocative take on the temporality of life (Owensby 1994).

8 Klaus Christian Köhnke has highlighted the authoritarian and anti-democratic undertones in Windelband's rejection of relativism (Köhnke 1986, 416–427). Perhaps an additional factor to take into account here is the institutional rivalry with his contemporary Dilthey.

9 It is only in his later writings that Dilthey puts forward *explicit* pronouncements regarding relativism. This raises the question as to whether these statements are responses to relativist tendencies in his earlier views. I do not have the space to argue for this point here, but I do not think this is the case. In my view, Dilthey's views from the 1890s are no less universalistic than those from the 1900s. When it comes to relativism, his mature and his late work stand and fall together.

10 Dilthey and Windelband's philosophies have undergone a series of shifts and transformations in the period from the early 1880s to their respective deaths in 1911 and 1915. I cannot give a diachronic account of their respective intellectual journeys in this chapter. But let me note that the conflict between their views is more salient in their writings from the 1880s and 1890s, while the convergence in views becomes more pronounced in their writings from the first decade of the twentieth century.

11   Husserl (1911) famously criticized Dilthey's philosophy as a brand of historicism that leads to "extreme sceptic subjectivism" (Husserl 1911, 323–326). While the theory of world-views indeed faces problems of relativism, it also aims to identify ahistorical essences. This calls into question whether its characterization as a form of historicism is adequate.

References

Bambach, C. (1995), *Heidegger, Dilthey and the Crisis of Historicism*, Ithaca, London: Cornell University Press.

Beiser F. (2008), "Historicism and neo-Kantianism," *Studies in History and Philosophy of Science* 39: 554–564.

—— (2014), *The Genesis of Neo-Kantianism, 1796–1880*, Oxford: Oxford University Press.

—— (2015), *The German Historicist Tradition*, Oxford: Oxford University Press.

Dilthey, W. (1880–1890), "Drafts for Volume II of the Introduction to the Human Sciences. Book Four," in Dilthey, *Selected Works I*, edited by R. Makkreel and F. Rodi, translated by J. Narnouw and F. Schreiner, Princeton: Princeton University Press, 1989, 243–391.

—— (1883), "Introduction to the Human Sciences," in Dilthey, *Selected Works I*, edited by R. Makkreel and F. Rodi, translated by M. Neville, Princeton: Princeton University Press, 1989, 47–240.

—— (1894), "Ideas for a Descriptive and Analytic Psychology," in Dilthey, *Selected Works II*, edited by R. Makkreel and F. Rodi, translated by R. Makkreel and D. Moore, Princeton: Princeton University Press, 2010, 115–210.

—— (1895–1896), "Contributions to the Study of Individuality," in Dilthey, *Selected Works II*, edited by R. Makkreel and F. Rodi, translated by E. Waniek, Princeton: Princeton University Press, 211–284.

—— (1903), "Rede zum 70. Geburtstag," in Dilthey, *Gesammelte Schriften V*, edited by G. Karlfried, Göttingen: Vandenhoek & Ruprecht, 1990, 7–9.

—— (1907), *The Essence of Philosophy*, translated by S. A. Emery and W. T. Emery, Chapel Hill: University of North Carolina Press, 1954.

—— (1910), "The Formation of the Historical World in the Human Sciences," in *Dilthey, Selected Works III*, edited by R. Makkreel and F. Rodi, translated by R. Makkreel and J. Scanlon Princeton: Princeton University Press, 2002, 101–212.

—— (1960a), "Das geschichtliche Bewusstsein und die Weltanschauungen," in Dilthey, *Gesammelte Schriften VIII*, edited by G. Karlfried, Göttingen: Vandenhoek & Ruprecht, 1–71.

—— (1960b), "Die Typen der Weltanschauung und ihre Ausbildung in den metaphysischen Systemen," in Dilthey, *Gesammelte Schriften VIII*, edited by G. Karlfried, Göttingen: Vandenhoek & Ruprecht, 73–118.

Ermarth, M. (1975), *Wilhelm Dilthey: The Critique of Historical Reason*, Chicago and London: University of Chicago Press.

Geldsetzer, L. (1968), *Die Philosophie der Philosophiegeschichte im 19. Jahrhundert: Zur Wissenschaftstheorie der Philosophiegeschichtsschreibung und -betrachtung*, Meisenheim am Glan: Hain.

Hartung, G. (2015), "Philosophical Historiography in the 19th Century: A provisional typology," in *From Hegel to Windelband: Historiography of Philosophy in the 19th Century*, edited by G. Hartung and V. Pluder, Berlin: De Gruyter, 9–24.

Heussi, K. (1932), *Die Krisis des Historismus*, Tübingen: Mohr.

Husserl, E. (1911), "Philosophie als strenge Wissenschaft," *Logos. Zeitschrift für Philosophie der Kultur* 1: 289–341.

Ineichen, H. (1975), *Erkenntnistheorie und gesellschaftlich-geschichtliche Wirklichkeit: Diltheys Logik der Geisteswissenschaften*, Frankfurt am Main: Klostermann.

Kinzel, K. (2017), "Wilhelm Windelband and the problem of relativism," *British Journal for the History of Philosophy* 25: 84–107.

Köhnke K. (1986), *Entstehung und Aufstieg des Neukantianismus: Die deutsche Universitätsphilosophie zwischen Idealismus und Positivismus*, Frankfurt am Main: Suhrkamp.

Kreiter E. (2002), "Philosophy and the Problem of History. Hegel and Windelband," in *Der Neukantianismus und das Erbe des deutschen Idealismus: Die philosophische Methode*, edited by C. Krijnen and D. Pätzold, Würzburg: Koenigshausen & Neumann, 147–160.

Lessing, H. (2016), *Die Autonomie der Geisteswissenschaften: Studien zur Philosophie Wilhelm Diltheys Bd. 2*, Nordhausen: Bautz.

Makkreel, R. (1975), *Dilthey: Philosopher of the Human Studies*, Princeton: Princeton University Press.

Oexle, O. (ed.) (2007), *Krise des Historismus–Krise der Wirklichkeit: Wissenschaft, Kunst und Literatur, 1880–1932*, Göttingen: Vandenhoeck & Ruprecht.

Owensby, J. (1994), *Dilthey and the Narrative of History*, Ithaca: Cornell University Press.

Patton, L. (2015), "Methodology of the Sciences," in *The Oxford Handbook of German Philosophy in the Nineteenth Century*, edited by M. Forster and K. Gjesdal, Oxford: Oxford University Press, 594–606.

Ranke, L. (1831–1832), "Idee der Universalhistorie," in Ranke, *Aus Werk und Nachlass IV*, edited by V. Dotterweich and P. Fuchs, München and Wien: Oldenburg, 1975, 72–89.

Rickert, H. (1921), *Der Gegenstand der Erkenntnis: Eine Einführung in die Transzendentalphilosophie*, Tübingen: Mohr.

—— (1924), *Die Probleme der Geschichtsphilosophie: Eine Einführung*, Heidelberg: Winter.

—— (1929), *Die Grenzen der naturwissenschaftlichen Begriffsbildung: Eine logische Einleitung in die historischen Wissenschaften*, Tübingen: Mohr.

Rodi, F. (2003), *Das strukturierte Ganze: Studien zum Werk von Wilhelm Dilthey*, Weilerswist: Velbrück.

Schnädelbach, H. (1984), *Philosophy in Germany, 1831–1933*, Cambridge: Cambridge University Press.

Steizinger, J. (2017), "Reorientations of Philosophy in the Age of History. Nietzsche's Gesture of Radical Break and Dilthey's Traditionalism," *Studia philosophica* 76: 223–244.

Troeltsch, E. (1922a), *Der Historismus und seine Probleme: Erstes Buch Bd. 2*, Berlin and New York: De Gruyter, 2008.

Troeltsch, E. (1922b), "Die Krisis des Historismus," *Die Neue Rundschau* 33: 572–590.

Windelband, W. (1882a), "Normen und Naturgesetze," in Windelband, *Präludien: Aufsätze und Reden zur Philosophie und ihrer Geschichte Bd. 2*, Tübingen: Mohr, 1924, 59–98.

—— (1882b), "Was ist Philosophie?," in Windelband, *Präludien: Aufsätze und Reden zur Philosophie und ihrer Geschichte Bd. 2*, Tübingen: Mohr, 1924, 1–54.

—— (1883), "Kritische oder genetische Methode?" in Windelband, *Präludien: Aufsätze und Reden zur Philosophie und ihrer Geschichte Bd. 1*, Tübingen: Mohr, 1924, 99–135.

—— (1891), *Lehrbuch der Geschichte der Philosophie*, Tübingen: Mohr, 1912.

—— (1894), "Geschichte und Naturwissenschaft," in Windelband, *Präludien: Aufsätze und Reden zur Philosophie und ihrer Geschichte Bd. 2*, 1924, 136–160.

—— (1905), "Geschichte der Philosophie," in Windelband (ed.): *Die Philosophie zu Beginn des zwanzigsten Jahrhunderts: Festschrift für Kuno Fischer Bd. 2*, Heidelberg: Winter, 175–199.

—— (1907), "Über die gegenwärtige Lage und Aufgabe der Philosophie," in Windelband, *Präludien: Aufsätze und Reden zur Philosophie und ihrer Geschichte Bd. 2*, Tübingen: Mohr, 1924, 1–23.

—— (1914), *Einleitung in die Philosophie*, Tübingen: Mohr.

Wittkau, A. (1992), *Historismus: Zur Geschichte des Begriffs und des Problems*, Göttingen: Vandenhoeck & Ruprecht.

Ziche, P. (2015), "Indecisionism and Anti-relativism. Wilhelm Windelband as a philosophical historiographer of philosophy," *From Hegel to Windelband: Historiography of Philosophy in the 19th Century*, edited by G. Hartung and V. Pluder, Berlin: De Gruyter, 207–226.

3   Hermeneutic responses to relativism

Gadamer and the historicist tradition

Kristin Gjesdal

Introduction

In Hans-Georg Gadamer's 1960 *Truth and Method*, the discussion of histori-cism figures at large. In Gadamer's view, historicism represents a response—an inadequate, even damaging response—to relativism. This response, he argues, is rooted in a misunderstanding of the relationship between the human and the natural sciences. Even though it is located more than a cen-tury back in time, historicism, in Gadamer's view, still shapes and limits her-meneutic philosophy. Hermeneutic philosophy is, in other words, caught in a picture; philosophers have gotten used to thinking that relativism can be dealt with in a particular way and that the label, marking this particular way, is that of historicism. From this point of view, a genuine response to the shortcomings of contemporary hermeneutics has to involve an effort to tackle historicism and provide a healthier, non-historicist answer to the perceived problem of relativism. In *Truth and Method*, Gadamer offers such a response by suggesting an approach that views the existence of a con-tinuous tradition as entirely fundamental to our self-understanding as well as our understanding of others.

Gadamer's *Truth and Method* has set the agenda for contemporary her-meneutic philosophy and even philosophy beyond a narrow hermeneutic tradition. After his initial debates with Jürgen Habermas and Karl-Otto Apel, Gadamer's position has influenced Charles Taylor, Alasdair MacIntyre, Richard Rorty, and the late Donald Davidson, and, more recently, Pittsburgh-based philosophers such as John McDowell and Robert Brandom.

The purpose of this chapter is to ask how Gadamer, in *Truth and Method*, approaches the challenge of relativism in historical thought and how he responds to the issues brought up by the constellation of philosophers that he labels historicists. The intuition I am working off is that Gadamer misunderstands both historicism and the relativist challenges to which it responds. This double misunderstanding will shape his hermeneutic contribu-tion in unwanted ways (including his notions of tradition and second nature to which Rorty, McDowell, Brandom, and other anglophone philosophers have recently turned).

Relativist puzzles

We first need to ask how Gadamer understands the problem of relativism in hermeneutics, i.e., relativism as it was faced by eighteenth- and early nineteenth-century philosophers who were thinking about understanding and interpretation along the lines of historicist thought. This question is the very starting point of *Truth and Method*. Gadamer, though, does not offer a clear analysis, definition, or exposition of relativism and historicism. Instead, he provides, in *Truth and Method*, a historical reconstruction that starts with Kant and the *Critique of the Power of Judgment*, moves to romantic philosophy, and only *then*, through what he pitches as the influence of misguided romantic repositories in philosophy, reaches Wilhelm Dilthey and the historicists. Yet, working our way backward, from the second to the first part of Gadamer's three-part opus magnum, we get a fair sense of what he has in mind. In the eighteenth century, what we today would speak of as the "human sciences"—and this is what Gadamer is mostly interested in—are under pressure from a number of sides.

Gadamer first identifies an early modern experience, across the sciences, the arts, and philosophy, of the modern mindset being constitutively different from the mindsets that dominated earlier periods. Then there is the notion, developed by Johann Gottfried Herder and other German eighteenth-century philosophers who were eagerly studying Hume and empiricist thought, that each historical period and culture is built around a set of values, concepts, beliefs, and practices that is unique to its time and must, as such, be understood on its own terms.[1] From this, it follows that if their concepts and values differ from ours—differ constitutively from ours, even—it is not, for that reason, a given that our concepts and values are more correct or more valuable than theirs are.[2] It is no matter of sheer coincidence that Herder is often listed as the father of modern anthropology (Forster 2010, 199–244). Nor is it an accident that he, in his time, was a staunch critic of slavery and colonialism.[3] Now, the question that will preoccupy hermeneutically minded philosophers—that will, in a certain way, give rise to hermeneutics in its modern form—is how we, given the possibility of such profound and constitutive differences, can at all understand others (and, relatedly, *know* that we understand others). What we face is, in other words, a problem of relativism: a worry that standards and values (that we can understand and apply) are relative to a given cultural horizon and that no meaning can be given to standards that transcend the given horizon or tradition at any given time. As constructed by Gadamer, historicism takes this worry seriously, and it seeks to come up with precisely such a standard (see, e.g., Gadamer 2003, 241; 1990, 245–246).

Gadamer, too, is interested in the problem of relativism. Yet, he does not, like the historicist models he discusses, recognize relativism as a genuine problem. Rather, he deems the problem of relativism a question that is unduly skeptical in nature. (He reasons that phenomenologically speaking, we do, after all, understand others most of the time and that the task of hermeneutics is to unpack and explicate the ontological structures of this basic understanding.) Eighteenth and nineteenth-century philosophy did not resolve this kind of skeptical anxiety.

On Gadamer's story, Kant's Copernican turn is a symptomatic and rather consequential signpost on the way to historicism. As Kant worried about the threat of historical and cultural variation, he sought to abstract from history and culture altogether by clarifying the transcendental conditions for valid judgments within the areas of epistemology, ethics, and aesthetic judgment. Gadamer takes the philosophical conditions of modern hermeneutics to be laid down at the point at which Kant, with the third Critique and its emphasis on the transcendental conditions for subjective-universal validity in the realm of taste (reflective judgment), brings an end to a long tradition of European humanism and its sense that value judgments, historical culture, and tradition are the domains in and through which human nature, second nature (culture and taste included), gets acted out and realized (Gadamer 2003, 40–41; 1990, 46).

Gadamer argues that well beyond the scope of Kantian philosophy proper, a Kantian mindset (broadly conceived) will mark hermeneutics and historicism in three ways. Firstly, it is assumed that the goal of the human-historical sciences is to reconstruct historical meaning objectively. Secondly, it is argued, along with a new orientation toward questions *quid juris*, that this goal is reached through adherence to an objectivity-establishing methodology (condition to meticulous application of the correct method, the correct results will follow in due course). Thirdly, it is claimed—not necessarily, but *de facto*, as a result of romantic currents in post-Kantian thought—that this methodology, as it addresses expressions of meaning, practice, and culture, entails an irreducible aesthetic component: a capacity to feel one's way into the minds or expressions of an other. For Gadamer, the historicist-hermeneutical paradigm is thus marked by a particular mix of positivism and aestheticism.[4] This is a trend that he finds in the works of Schleiermacher, Dilthey, and others. At stake, as Gadamer views it, is a set of positions that, while recognizing the challenge of historical change and the (relative) autonomy of cultural paradigms, responds by hypostatizing an aesthetic-scientific subjectivity that is somehow independent of, or detached from, history and, by virtue of these qualities, can traverse historical and cultural distance, feel its way into, and (quasi-scientifically) reconstruct historical meaning (Gadamer 2003, 164–169; 1990, 169–174). On this model, the problematic relativity of historical horizons is, in other words, overcome by reference to a trans-historical, transcendentally grounded standard that, at the end of the day, is led back to the interpreting subjectivity (understood along the lines of a disinterested scientist) and not to the negation of historical difference in and through the vicissitudes of a developing tradition of which the interpreter is a part (Gadamer 2003 165; 1990, 170).

In Gadamer's view, this is an untenable position. Historicism is based on a series of fundamental contradictions. While it seeks, for a start, to obtain historically neutral judgments on historical material, while it seeks to over-come the historicity of the interpreter by reference to a set of proto-scientific, methodological guidelines, the historicist paradigm Gadamer criticizes ends up hypostatizing—as a trans-historical and allegedly neutral epistemic point of departure—the values of its own post-Kantian outlook. (Sometimes, Gadamer even speaks of a post-Cartesian outlook; see, e.g., Gadamer 2003, 277–278, 461; 1990, 281–282, 465.) That is, in the historical sciences, the image of a neutral and ahistorical scholar whose rationality rests with the ability to arrive at rules and concepts that are thought to be universal and necessary, i.e., above and beyond historical change and variation, is itself a historically and culturally tainted conception.

There is, for Gadamer, no such thing as a "neutral" or an "objective" approach to history, at least not if objectivity is thought to imply a view from above, independent of history and culture. While historicism seeks an ordered and formally conceived rationality—mapping and making sense of the messy landscape of past human thoughts and actions—it ends up basing its judgments on a model that, for Gadamer, is unacceptable. It is both unduly *subjectivizing* in that it is rooted in the individual interpreter's ability to feel her way into the meaning of the historical material at stake and unduly *object-ivizing* in that this feeling subjectivity is conceived as elevating itself above his-tory and culture and thus as studying an object with which it, pre-reflectively and phenomenologically speaking, has no deeper affinity. Finally, Gadamer attacks the very idea of historical reconstruction, i.e., the goal in light of which this model of history is conceived. Historicism sought to bring history closer to us. It sought to take history seriously as a field of scholarship and to show why it really matters to finite beings of our kind. In reality, Gadamer argues, the movement represented a hollowing out of history—a reduction of history from being a general horizon of cultural self-understanding and practice to its being a special branch of academic scholarship. As a result, historicism ultimately marginalized our relationship to history and tradition. The past is now seen as irrevocably past (rather than as a deeper dimension of meaning that shapes the horizon of the present). Regardless of how much his-tory is infused with feeling, it will, on this paradigm, end up as museum-like, dusty, and with little or no relevance for the present.

For Gadamer, this lack of mediation between history and present is a particularly pressing problem in the realm of art. Thus, parts one and three of *Truth and Method* concentrate on art and aesthetic experience. Without the appeal to tradition, without a sense of history, as a field of meaning and interest, we are deprived of a larger sphere through which human self-formation and -understanding can take place.[5] The past is objectivized, distanced, and disengaged from the present, and as a result, we are stuck with an unnecessary (and harmful) scientistic approach to history but also with a massive existential vacuum. (How to make sense of our lives in history and culture if history and culture can no longer function as an arena of self-understanding?) Hence, from Gadamer's point of view, action is required, and it must involve an effort to liberate hermeneutics from the shackles of historicism.

Retrieving tradition

For those of us who are not card-carrying Gadamerians, it seems obvious that Gadamer is too polemical in his description of historicism and the cluster of hermeneutic positions that, in his view, adhere to a historicist paradigm broadly conceived. That is, *if* historicism were the kind of naïve, antiquarian, subjectivist, and formalist blend that Gadamer makes it to be, *then* he might well be right in seeking to dismantle this paradigm. But that, in my view, is very far from what historicism is. (For solid discussions of historicism, see Bambach 1995 and Beiser 2012.) But let us bracket this point for now. (I will briefly return to it below.) For against his criticism of historicism, past and present, Gadamer articulates his own, alternative model of the nature and task of hermeneutics, i.e., philosophy of understanding and interpretation. It is to this model that I now turn, asking whether and how Gadamer's own philosophy is shaped by the philosophical problems and challenges it is designed to overcome.

The main argument of *Truth and Method* is reasonably straightforward. Gadamer points out that if we, as practitioners of human-historical science, think that we are above history, if we, as historical creatures, think that our judgments are beyond or independent of history and the way it shapes human thought, then we will eventually be more bound by history than necessary— or, rather, we risk being bound by history in negative ways only. (At the very least, we risk being bound up with it *also* in negative ways; see Gadamer 1990, 286–287.)

In historicizing the object interpreted, in placing it within a cultural paradigm that is long gone and thus subject to objectification, the historicist models that Gadamer criticized had sought to overcome the cultural situatedness of the interpreter (so that she could bracket her own historical horizon and reach an other, constitutively different horizon of meaning). They had, as it were, hypostatized the image of an interpreter whose gaze is beyond, and thereby untainted by, historically and culturally coined concepts, beliefs, and prejudices. But such a point of view, Gadamer argues, is simply not available to finite, human beings. We are, emphatically, historically and culturally situated creatures. If we deny the power of prejudices, we risk being subject to them in naïve and potentially harmful ways.

It must be kept in mind at this point that for Gadamer, "prejudice" is a value-neutral term, simply another name for a human-historical situatedness; it refers to pre-judgments, the basic background against which judgments are enabled, and, as such, something that is not yet subject to reflective checkup and judgment (Gadamer 2003, 306; 1990, 311). Prejudices can be negative but need not be. In either case, though, what is needed is a turn *from* the ideal of an interpreter bracketing or abstracting from prejudices, *to* an attempt to thematize them, make us aware of their influence, and respond to them in a mature and productive (but not, for that reason, uncritical) way. Only by acknowledging the power of prejudices, only by bringing them into the open and into the space of reason, can we consider their relevance and validity. Designated as a pre-reflective web of beliefs, prejudices can be productive as well as limiting. But hidden or unacknowledged prejudices have more power over us than prejudices whose existence is recognized and brought into a space that allows for critical examination and active appropriation.

At this point, Gadamer's position gives rise to an obvious question. If history is the horizon from within human reflection and reasoning takes place, how, if at all, can we get around (or, again, *know* that we are getting around) the prejudices bestowed on us by history and make progress toward a better understanding? For Gadamer, this is precisely the kind of question that had originally led the historicists astray. It is the wrong kind of question and reflective of the wrong kind of historical understanding, thus easily evoking the wrong kind of hermeneutic responses. For historical work, on Gadamer's understanding, is not about isolated horizons or a set of beliefs and/or prejudices, be they those of the interpreter or those reflected in the historical expression. Rather, historical-hermeneutic work implies a *fusion of horizons* (Gadamer 2003, 306; 1990, 311). A fusion of horizons, as Gadamer sees it, both presupposes and adds to the continuum of tradition. It is through such a concept of tradition that Gadamer seeks to overcome the historicist challenge to relativism. For if horizons, against the continuum of tradition, can merge and expand, then the interpreter is *not* locked into a local historical context but can move toward a dialectical-dialogical encounter with a text from another context (but within the tradition, which, for Gadamer, is basically one).

In engaging expressions from a distant historical period, an interpreter encounters both a part of her tradition *and* a different way of thinking. This, in turn, makes her reflect on her own horizon and the values, beliefs, and concepts that make up its building blocks; it makes her, in short, realize that her horizon, too, is constitutively historical and that, even with the aid of a zealous, scientific consciousness, she cannot leap out of tradition so as to embark on some sort of value-neutral and methodologically purged enterprise of hermeneutic-historical reconstruction. From this point of view, the problem with historicism is its effort to overcome relativism by seeking to get beyond our situatedness in history, its failing to see that tradition itself can serve as a source for true understanding.[6] The problem is not, as such, the particular way in which historicism seeks to overcome historical situatedness and distance (i.e., the appeal to methodological consciousness). Rather, it is the very goal, the philosophical ambitions, of historicism that turns out to be wrong from the outset. Neither historical distance nor the historical situatedness by which it is conditioned is a problem in need of a solution. Quite the contrary, for Gadamer, historical distance discloses a possibility to thematize, foreground, and also expand the boundaries of one's

own horizon: not by leaping out of it but by widening it through the confrontation of other ways of thinking and acting. A horizon, for Gadamer, has
fussy—dynamical and flexible—borders (see, e.g., Gadamer 2003, 302–206;
1990, 307–311). It can be subject to ongoing revision, expansion, and change.
The ability to engage in encounters that facilitate such expansion and change
is, for Gadamer, a hallmark of human rationality and, what is more, it is an
activity that rests at the core of the human-historical sciences.

The human-historical sciences, Gadamer argues, are dialogically based,
and as such, they facilitate productive hermeneutic-historical encounters.
Further, this is one of the things that, if not necessary and universally (though
Gadamer at times seems to think that it is so), then at least often and typically set scholarship in these areas apart from the pursuits of the natural
sciences: At stake is a process of learning, and the learning we here encounter
is *both* about the object under investigation *and* the interpreter, who, as a
human-historical being, is herself a product of the history/tradition under
investigation. On Gadamer's model, the continuum of tradition and language means that it is virtually impossible to understand an expression from
another culture or time period without also highlighting, thematizing, and
possibly correcting and/or expanding one's own horizon of understanding. In
this sense, understanding always involves a dimension of self-understanding,
although for historical creatures, this self-understanding can never be complete (Gadamer 2003, 302; 1990, 307). And, vice-versa, self-understanding
grows in and through the kind of interpretative work that the humanities
facilitate. Gadamer terms this process *Bildung*, sometimes translated as education in culture and history.[7]

For Gadamer, history is conceived of as tradition—and tradition, in turn,
as an unbroken conversation between generations. Thus, the past is not, as
the historicists had assumed, alien to us: there is no (constitutive) distance
to be (epistemologically) overcome, and thus also no relativist problem
to worry about. To the extent that we think of the past in terms of human
practice, expressions, and horizon-shaping works, history is passed down
in and through centuries of reading, interpreting, and other kinds of mediation. Works of philosophy, literature, and art are enabled by a tradition in
which the interpreter, assuming she is of a background to which these works
matter, is herself a part. Hence, in turning to the works of tradition, what
the interpreter faces is not an object that is distant and that presents her, like
the historicists had argued (on Gadamer's reading), with a historical gap or
relativist challenge that needs to be traversed or overcome. No, what happens
here, rather, is that in encountering these works, she encounters, in a semi-
metaphorical but nonetheless quite serious sense, herself.

For Gadamer, there is, at the end of the day, no problem of relativism.
Instead, we face a thick, productive evolution of historical meaning, one in
which certain aspects of our outlook on the world will be brought to the
foreground and highlighted. There will certainly be elements that we cannot
understand, things that will surprise and baffle us. However, emerging against

a background of continuity, elements of relative incomprehension will turn us, as interpreters, back on ourselves and have us reflect on and get a better grasp on our own historicity and our own horizon. Stronger still, being historical creatures, we *only* get a grasp on our own historicity in and through encounters with other historical horizons.

When viewed in this way, the productivity of historical-hermeneutic work is not a matter of scientific reconstruction or overcoming temporal distance and cultural difference, but of learning who we are. Tradition is the medium in which an interpreter lives and thinks. It shapes her before she can reflect on and take a stance toward it. She can only grasp it by acknowledging its influence and how it discloses the world to her *as* a world of human significance.

If there is, in this way, no threat of a fundamental and deeply entrenched relativity between cultures, then we no longer need the methodological machinery that the historicists, on Gadamer's reading, had presented us with. Instead of viewing historical-hermeneutical work in terms of what guidelines an interpreter should actively adopt in order objectively to reconstruct the meaning contexts or events of the past, instead of seeing historical-hermeneutic endeavors as a less precise variety of laboratory work, we need to recognize that there is an element of interest at stake, and this element is not a weakness but rather a condition of possibility for (and possible strength of) this kind of scholarship. This, in short, is Gadamer's argument as he seeks to overcome what he takes to be the misguided orientation of historicism.

Critique revalued

In my view, the goals of Gadamer's hermeneutics are mostly sound: As I read his work, he takes hermeneutics to involve a genuine effort to put prejudices to a test and recognizes that in order to understand an other, an interpreter often has to revise, modify, or even change her beliefs and assumptions, be they pre-reflective or reflectively held. Further, I appreciate Gadamer's suggestion that historical understanding is one of the ways in which finite, historical creatures, creatures like us, can thematize pre-reflective prejudices and take a stance on them. As historical creatures, we have no available point of nowhere, yet we can, on this view, make productive use of our points from somewhere. In this way, hermeneutic encounters facilitate a process of learning: learning about others, learning about the world (as thematized in and through expressions and practices), and learning, in a stronger or weaker sense, about oneself. All three components have to be there for genuine (historical) understanding to take place. Thus understood, we can address this as a process of *Bildung*.

However, accepting this general approach to understanding and *Bildung* is not the same as to accept Gadamer's understanding of tradition as facilitating an overcoming—or undermining—of the relativist problem that had, originally, triggered the historicist mindset. Nor is it the same as to accept the particular ontology of understanding that undergirds this Gadamerian approach.

In order to understand a text from the past, in order to overcome the methodological-objectivizing attitudes that have shaped our modern (historicist) approach to hermeneutical work, Gadamer recommends that we obtain a role in which we "listen" to history and allow tradition to present itself to us as the context (horizon) in which we stand. Our relationship to tradition, he suggests, is that of a play-like interaction. And just like it is a game, it is not the individual player but the play/tradition itself that is constitutive for the participation of the individual interpreter or player (Gadamer 2003, 102–104; 1990, 108–110). In this sense, playing is a mix of passivity (one subjects oneself to the rules of the play) and activity (the rules still have to be applied, and there is room for spontaneity and self-expression). Yet, it is the passivity involved that actually enables activity. At stake is an ability to apply the (shared and passively absorbed) meaning of the work in the (particular) context of the interpreter. As it is, understanding, for Gadamer, depends on this element of application. We are only able to make sense of the meaning of the expressions and events of the past if or when we can apply it and make sense of it within our own context of understanding (Gadamer 2003, 309; 1990, 314, see also 297, 302).

Given Gadamer's emphasis on the continuity of the tradition and the passivity of the interpreter in "listening" to the voices of the past, the question I would like to ask in this context is whether his philosophy—emphasizing, in a strong sense, the continuity of tradition and, relatedly, collapsing the distinction between interpretation and application—really offers the best tools for thinking about understanding as a process of *Bildung*. In my view, it does not and, further, this has to do, exactly, with how he polemically constructs and responds to his historicist opponent (i.e., the problem that his contribution was designed to overcome). For the very point that gets Gadamer's hermeneutics going is his insistence not only on the continuity of the tradition, but also, and closely relatedly, on the meaning of historical works, expressions, and events being made available to us because they have been meditated in and through a continuous tradition.[8]

There are cases where such a point *prima facie* makes sense. Those cases sometimes (but not always) relate to the interpretation of art and, even more so, the meaning of canonical works—the works that are and have been taken to be constitutive for a given culture and its dominant models of self-understanding. By reference to works such as William Shakespeare's *Hamlet* or Henrik Ibsen's *A Doll's House* or, further back in time, a tragedy by Sophocles, it would not be outrageous to suggest that these works are available to us precisely because readers, past and present, have cared about them. Their meaning transcends the possible intention of their erstwhile authors; they have accumulated a hermeneutic patina, an excess of meaning, that makes them, as it were, more than pieces of literature written by individual authors in given cultures at a given point in time. If we were to approach *Hamlet* without an understanding of how Shakespeare's play has shaped modern theater and the modern understanding of melancholy and interiority, we would miss out on a significant dimension of the work and of literature and drama, perhaps even Western culture. Likewise, Ibsen's *A Doll's House* came to be (and still is) a literary beacon of gender politics and, in its time, contributed to the shaping of the new woman—on and off stage—and to an entirely new way of producing and thinking about theater. Gadamer, in my view, is right in emphasizing the historical meaning of art—how an artwork's meaning accumulates, gets thicker, over time—over and against the kinds of historicism that locate the meaning of a work solely within its context of origin. However, where Gadamer is wrong, as I see it, is in that he does not allow for a clear distinction between the descriptive and the normative aspects of this mediation in tradition. That is, he assumes, in a spirit that is far too conservative for my philosophical compass, that tradition is self-repairing, self-correcting, and somehow will facilitate an ever richer and more productive meaning.

During the debates that followed the publication of *Truth and Method*, Gadamer's critics pointed out that his model fails to account for the negative aspects of prejudices: that he had a tendency to hypostatize the role of pre-reflective belief-systems as they enable and disclose a world of meaning and that, as a result, he fails to consider how tradition can also lead us astray and cement existing systems of power, domination, and ideology. Now, in my view, Gadamer, while not saying much about the negative impact of prejudices, does not categorically rule out the possibility that prejudices can be and indeed often are negative. His point, as I read him, is rather to emphasize, against what he views as a tendency, in the wake of historicism, to postulate the ideal of a neutral and, so to speak, ahistorical interpreter, the historicity of historical understanding (Gadamer 2003, 309; 1990, 314, this is further emphasized in his discussion of the classical, see, e.g., 288, 292–293). He wants to see historical understanding as a process in which false prejudices are revised in and through the hermeneutic process.

Gadamer's claim is not that this always happens but that, at its best, this is something that understanding, in the human sciences, can achieve and that, as such, it should inform hermeneutic philosophy. Gadamer is right—and genuinely original—in his emphasis on the productivity of tradition. Where I do not follow him is in his trust that the unfolding of tradition is itself sufficiently powerful (sufficiently rational) to facilitate a process of reflection and correction: that there is a healthy productivity built into tradition, understood as a series of hermeneutic encounters, that gives rise to an ongoing questioning of prejudices. It is, I fear, naïve to think of this as a result of an interpreter's ability to listen, be played, and remain passive in the encounter with the tradition. I worry that this is but a Heideggerian residue in Gadamer's work, and, moreover, a dimension of it that should be questioned. It is far from evident that passivity, listening or opening oneself up to the meaning of a past text, work, or even part of the tradition, will offer the best resources to overcome unhealthy prejudices. In order to overcome unhealthy prejudices and bias, we need reflection, criticism, and a commitment to philosophical genealogy, to put it in Nietzsche's terms. That is, we need a vocabulary that can help us, without an appeal to a Kantian *a priori*,[9] to determine what prejudices need to be overcome in the first place, how they got a hold on us, and how historical-cultural work, if at all, can help facilitate strategies of demystification and enlightenment. In addition to appealing to the resources of critical theory, hermeneuticians could do well in drawing on the recourses of later, anglophone philosophy. It would, for example, be helpful to bonds with social epistemology and the work that is recently being done on bias, hermeneutic injustice, and related issues.[10]

Given the tendency of tradition to (auto-)confirm certain values, works, or aspects of works as good, valid, and worthy of attention and others as less so, we must ask how an interpreter can know that what she reads and cares about, precisely, are works that can challenge and expand her present self-understanding and invite a revision of her prejudices and bias. Gadamer's philosophy offers few resources here. Further, his emphasis on the continuity of tradition (and the auto-productivity it fosters) does not hold up to scrutiny at the empirical-historical level.

There is no shortage of examples of how a tradition has been revised, its basic beliefs and criteria questioned when a group, often an under-represented group (whose expressions have been marginalized), casts doubt on the continuous and dominant meaning of the canon. Just think of the recent work done by early modern scholars to include women philosophers into their canon. This is not simply a matter of adding a new group of philosophical text (quantitatively speaking) but also an effort to change our understanding of the period and the systematic-philosophical questions that emerge within it (qualitatively speaking). And did not, in a similar manner, the inclusion of the Harlem Renaissance into the U.S. canon also alter our understanding of modernism in American literature, music, and painting? Should not these processes of historical making-right (which cannot be accounted for by appeal to the continuity of tradition alone) also be followed by reflection on the mechanisms of exclusion, bias, and cultural privilege that led to the relative suppression of marginalized voices or movements in the first place?

My point here is not that Gadamer (or a Gadamerian) would be forced to deny the relevance of such exercises. Rather, the real question, the hard, philosophical question, is how far the metaphors of passivity, listening, playing, and absorption can take us when it comes to critiquing the premises of our own tradition and the prejudices it shelters. Tradition is not free of ideology, but one of the ways, even a dominant way, in which ideology is cemented.

This is the point the historicists saw (and, again, I have in mind the historicists in the lineage from the young Herder to Dilthey) and what Gadamer, in reducing their efforts to a response of unhealthy aesthetic positivism in the face of a relativist threat, is prone to overlooking.[11] In order to retrieve this dimension of historicism, historicism in its promising *ur*-form, we need, however, to move back in time and return to a part of this legacy that is historically prior and systematically diverging from the post-Kantian philosophy that Gadamer criticizes. This early version of historicist thought is not prone to the kind of fallacy he worries about, but it does not, for that reason, offer a hermeneutic model that, like Gadamer's, is shaped by a commitment to a world-disclosive continuum of meaning and understanding.

Relativism reassessed

In Gadamer's view, historicism represents a reduction of tradition and historical meaning, from being, along the lines of a broader humanist model, a productive source of self-understanding and identity to representing a problem, a possible source of distance and difference, and something that needs to be overcome.

However, historicism is not and should not, as Gadamer does, be reduced to a post-Kantian phenomenon. It is, rather, a position that emerges in Kant's own time and as a *deliberate* alternative to transcendental philosophy and its insistence on philosophy as a methodological propaedeutic for real science. It grows out of a complex cluster of impulses—the rise of the biological sciences, a new interest in empiricist philosophy, an attempt to stay true to the ideals of philosophical induction, and a critical reaction to Eurocentric politics, colonialism, and slavery, as it was justified with reference to the superiority of one tradition over others.[12]

One example of this is the young Herder's reaction to ahistorical models of understanding, especially those following from rationalist school philosophy, and his frustration with the kind of stifling historical paradigms— *ahistorical* historical paradigms; the elevation of certain historical periods into a trans-historical ideal of culture as such[13]—that went hand in hand with the classicist paradigm of the day. Against this, Herder launches a historicist program that, in its early period, is oriented toward a thematizing of historical-cultural blindness and prejudices. His hermeneutics spans both historical understanding *and*, closely related, inter-cultural understanding. Like Gadamer, Herder emphasizes how a final historical reason, human reason, can only prosper and thrive to the extent that it is exposed to, reflects on, and takes seriously a plurality of possible outlooks on life and approaches to ethical, moral, and political questions. Yet, he does *not*, at least not as I read him, fall prey to relativism. Slavery, for instance, is universally abandoned (rather than being pitched as relative to cultures), likewise are widow burning and caste systems (Herder 1989, 456). In his early period, Herder is motivated by Enlightenment humanism and his belief in a human nature that will grow and develop from and through the exercise of tolerance and learning in the encounter with others.

When Gadamer seeks to replace the appeal to method with an appeal to truth, he unintentionally risks a backlash to a period prior to the Enlightenment agenda—a backlash that he himself flirts with in his wordplay on the German terms of *wahr* (true) and *bewahren* (preserve). The young Herder, by contrast, appeals to criticism and an insistence on the constant need to work toward an overcoming of prejudices—and the need to do so precisely in order to *preserve* truth or, rather, the broad commitment to objectivity that is a condition for our getting closer to truth and truthfulness in the human-historical sciences. However, in my view (and unlike Gadamer's), this is not an example of a naïve Enlightenment spirit but of a productive commitment to Enlightenment ideals, broadly conceived: a commitment to the Kantian *sapere aude* within the field of historical scholarship. And while Gadamer's developmental-descriptive narrative in *Truth and Method* hypostatizes the continuity and normative weight of tradition, contemporary hermeneutics should also revisit the early Enlightenment roots of the discipline and explicate—precisely by allowing us to question the ontological dominance in post-Heideggerian hermeneutics— the critical resources inherent to philosophical hermeneutics.

By way of conclusion, I would like to stress the point that historicism emerges out of an experience of, and response to, the problem of relativism. Unlike Gadamer, however, I take the historicist movement to represent a powerful response to the challenges of relativism—one in which the relative autonomy of cultures and periods is perceived as a genuine philosophical challenge *and* as potentially productive. It is, precisely, the difference, sometimes even incompatibility, between cultures and horizons that, for a finite, historical reason, facilitates growth, understanding, and *Bildung*. In overlooking this simple point, Gadamer overlooks a vital resource within the hermeneutic and historicist traditions—*and* one of the ways in which these traditions can offer helpful, philosophical resources also beyond a philosophy that is committed to a program of an ontological (or post-Heideggerian) kind.

Notes

1 Gadamer refers to Herder throughout *Truth and Method* but never engages his contribution in much detail. He discusses the relevance of Herder's philosophy in an essay (originally a lecture given to French prisoners of war) published during the war as *Volk und Geschichte im Denken Herders* (Gadamer 1942). An edited version of this essay was later included in Gadamer 1999a, 318–335.

2 As Herder famously puts it, each culture has its center of happiness in itself (Herder 2002, 299; 1994, 41). See also the extensive discussions in Spencer (2012) and Sikka (2011).

3 Herder points out that even though Europe has officially abandoned slavery, people in Europe still continue "to *use* as slaves, to *trade, to exile* into silver mines and sugar mills, three parts of the world." He further criticizes how French and German intellectuals assume that "when a storm shakes two smalls twigs in Europe ... the whole world quakes and bleeds" (Herder 2002, 238, 325; 1993, 73–74; 1994, 70).

4 See Gadamer 1985, 157–182. This essay draws on his more extended discussion in *Truth and Method* but also reverberates with Heidegger's concerns from *Being and Time*.

5 Here is Gadamer's undoubtedly conservative (and conservatory) description of the predicament with respect to art and culture: while, in the past, art "occupied

a legitimate place in the world, [so that] it was clearly able to effect an integration [*eine selbstverständliche Integration*] between community, society, and the Church, on the one hand, and the self-understanding of the creative artist, on the other ..., our modern problem, however, is precisely the fact that this self-evident integration, and the universally shared understanding of the artist's role that accompanies, no longer exists [*daß diese Selbstverständlichkeit und damit die Gemeinsamkeit eines umfassenden Selbstverständnisses nicht weiterbesteht*]" (Gadamer 1986, 19; 1999b, 110).

6 As indicated by the title of *Truth and Method*, there is, in other words, a shift from objectivity to truth, and truth, in turn, is broadly conceived in Gadamer's work.

7 This notion is later endorsed in Rorty (1980) and McDowell (1994) and (2002). For Gadamer's historical retrieval of this term, see, for example, Gadamer 2003, 35; 1990, 41).

8 For Gadamer, it is the continuity of tradition that legitimizes the emphasis on application. If it were not for the continuous (normative) background of tradition, the emphasis on application would itself be subjectivist. Arguably, one could portray this as a Hegelian component in Gadamer's work, though Gadamer questions Hegel's notion of absolute spirit—indeed sees his own hermeneutics as aspiring to fill "the place vacated by Hegel's absolute" (Gadamer 2003, 230; 1990, 234).

9 In Gadamer's work, there is a tendency to associate any critical-reflective approach with the tradition of a Kantian *a priori* and a naïve appeal to methodology. The tradition from Herder to Nietzsche, however, clearly demonstrates that this is wrong.

10 For the Gadamer-Habermas debate, see Warnke (1987) and Bernstein (1983). For the notion of hermeneutic injustice, see Fricker (2007).

11 Given Gadamer's existential-ontological focus, there is a tendency in his work to assume that all epistemology is bad (for him: quasi-Cartesian) epistemology.

12 For a fuller account, see Gjesdal 2017.

13 It should be noted how Herder here anticipates Nietzsche's approach in *The Genealogy of Morals* as it is articulated in his opening remarks and criticism of (English) moral psychology.

References

Bambach, C. (1995), *Heidegger, Dilthey, and the Crisis of Historicism,* Ithaca: Cornell University Press.

Beiser, F. (2012), *The German Historicist Tradition*, Oxford: Oxford University Press.

Bernstein, R. (1983), *Beyond Objectivism and Relativism: Science, Hermeneutics, and Praxis*, Philadelphia: University of Pennsylvania Press.

Forster, M. (2010), *After Herder: Philosophy of Language in the German Tradition,* Oxford: Oxford University Press.

Fricker, M. (2007), *Epistemic Injustice: Power and the Ethics of Knowing*, Oxford: Oxford University Press.

Gadamer, H.-G. (2003), *Truth and Method*, translated by J. Weinsheimer and D. G. Marshall, New York: Continuum.

—— (1990), *Gesammelte Werke 1*, Tübingen: J. C. B. Mohr.

—— (1999a), *Gesammelte Werke 4*, Tübingen: J. C. B. Mohr.

—— (1999b), *Gesammelte Werke 8*, Tübingen: J. C. B. Mohr.

—— (1986), *The Relevance of the Beautiful and Other Essays*, edited by R. Bernasconi, translated by N. Walker, Cambridge: Cambridge University Press.

—— (1985), "Wilhelm Dilthey nach 150 Jahren (Zwischen Romantik und Positivismus. Ein Diskussionsbeitrag)," in *Dilthey und die Philosophie der Gegenwart*, edited by E. W. Orth, Freiburg: Karl Alber, 157–182.

—— (1942), *Volk und Geschichte im Denken Herders*, Frankfurt am Main: Klostermann.

Gjesdal, K. (2017), *Herder's Hermeneutics: History, Poetry, Enlightenment*, Cambridge: Cambridge University Press.

Heidegger, M. (1996), *Being and Time*, translated by J. Stambaugh, Albany: SUNY Press.

Herder, J. G. (2002), *Philosophical Writings*, edited and translated by M. Forster, Cambridge: Cambridge University Press.

—— (1993), *Werke in zehn Bänden 2*, edited by E. Bollacher and G. Grimm, Frankfurt am Main: Deutscher Klassiker Verlag.

—— (1994), *Werke in zehn Bänden 4*, edited by E. Bollacher and J. Brummack, Frankfurt am Main: Deutscher Klassiker Verlag.

—— (1989), *Werke in zehn Bänden 6*, edited by E. Bollacher, Frankfurt am Main: Deutscher Klassiker Verlag.

McDowell, J. (1994), *Mind and World*, Cambridge, Mass.: Harvard University Press.

—— (2002), "Response to Rüdinger Bubner," in *Reading McDowell*, edited by N. Smith, London: Routledge, 296–297.

Nietzsche, F. (1967), *On the Genealogy of Morals*, translated by W. Kaufmann and R. J. Hollingdale, New York: Random House.

Rorty, R. (1980), *Philosophy and the Mirror of Nature*, Oxford: Blackwell.

Sikka, S. (2011), *Herder on Humanity and Cultural Difference: Enlightened Relativism*, Cambridge: Cambridge University Press.

Spencer, V. (2012), *Herder's Political Thought: A Study of Language, Culture, and Community*, Toronto: University of Toronto Press.

Warnke, G. (1987), *Gadamer: Hermeneutics, Tradition and Reason*, Stanford: Stanford University Press.

Part II

Science

Introduction

Martin Kusch

In this day and age, (cognitive) relativism is often regarded as anti-scientific. For instance, the main "whipping boy" of the 1990s "Science Wars" was the relativistic sociologist who (allegedly) sought to downgrade natural science by treating it as on a par with myth or magic.

Interestingly enough, this association between relativism and disrespect for science was not prominent in debates over relativism during the "long nineteenth century" in the German-speaking world—roughly, from Georg Wilhelm Friedrich Hegel (1770–1831) to Adolf Hitler (1889–1945). On the contrary, in this time period and geographical region, relativism was often seen as an *obvious consequence* of natural-scientific attitudes and theorizing. Philosophers were divided on how to assess this consequence. Some saw relativism as an essential element of the modern scientific worldview; others attacked relativism by portraying it as the result of what we today might call "scientism" or "scientific imperialism."

To understand this intellectual constellation, we need to remember four important features of the natural sciences in the long nineteenth century. First, the time period in question witnessed numerous scientific advances that were perceived as "revolutionary" by many contemporaries. Moreover, Charles Darwin (1809–1882), Ernst Haeckel (1834–1919), Hermann von Helmholtz (1821–1894), Bernhard Riemann (1826–1866), Wilhelm Wundt (1832–1920), Ernst Mach (1838–1916), or Albert Einstein (1879–1955)—to mention just a few—were scientists whose names were familiar to readers of highbrow newspapers and popular weeklies. This was due, in no small measure, to these scientists' own efforts in popularizing their findings.

Second, in the German-speaking world natural scientists, philosophers, historians, linguists, and economists still belonged to the same faculties. This made for close interactions across disciplinary boundaries: thus, one finds historians trying to learn from biology or psychology (e.g., Karl Lamprecht (1856–1915), Friedrich von Hellwald (1842–1892)); philosophers engaging closely with sense-physiology (e.g., Alois Riehl (1844–1924]); logicians and epistemologists seeking to integrate their investigations with the psychology or biology of reasoning (e.g., Benno Erdmann (1851–1921), Mach);

physiologists presenting their results in a Kantian garb (e.g., von Helmholtz); or physicists highly sensitive to epistemological debates (e.g., Einstein).

Third, during the long nineteenth century, the classification and institutional division of the sciences and humanities underwent substantial and often conflictual re-organization. The first chairs for physiology were introduced in the 1850s, and psychology began its long and painful separation from philosophy in the 1890s. Sometimes such re-organizations created "split identities," that is, authors who could claim to have substantive expertise in more than one field. Thus, Wundt was both a psychologist and a logician; von Helmholtz a physiologist, physicist and philosopher; or Mach a physicist, epistemologist, and psychologist. New institutional boundaries frequently led to fierce competition over academic chairs; for instance, universities and ministries of education tended to create chairs in experimental psychology by cutting positions in traditional fields of philosophy. The traditional philosophers ultimately responded with a petition.

Fourth, for much of the long nineteenth century, philosophy struggled to recapture the cultural capital it had lost when the systems of Friedrich Wilhelm Josef Schelling (1775–1854) and Hegel fell into disrepute. The struggle was difficult not least because some scientists (e.g., Haeckel) published widely circulating books in which they declared much of traditional idealistic philosophy obsolete.

The last four paragraphs can help us appreciate why in the long nineteenth century, relativism and natural science were seen as closely intertwined. To begin with, many important natural-scientific results were presented by their proponents as undermining traditional philosophical beliefs in absolutes: mathematicians showed that Euclidean geometry was not without alternatives; physicists rejected Newtonian conceptions of absolute space and time; biologists replaced eternal and immutable species with contingently evolving species; statisticians supplanted human essence with the fiction of *l'homme moyen*; sense-physiologists argued that perceptions are structured in good part by needs of the organism; and cognitive, developmental and social psychologists endlessly displayed different forms of "apperception," that is, the influence of background information on belief formation.

The scientific challenging of absolutes was sometimes influenced by earlier or contemporaneous work in philosophy or politics. John Stewart Mill (1806–1873) and Herbert Spencer (1820–1903) were particularly important here. But the scientific rejection of absolutes in turn also had a substantive impact on philosophy and stimulated forms of philosophical cognitive relativism. Clear cases in point were some of the logicians and epistemologists whom Gottlob Frege (1848–1925) and Edmund Husserl (1859–1938) would later attack as "psychologistic." One particularly striking self-proclaimed relativist was the philosopher Georg Simmel, who assembled his "relativistic worldview" out of scientific insights from Darwin to von Helmholtz, Riemann to Einstein. Simmel was a student of Berlin philosophers, early social psychologists (e.g., Moritz Lazarus), and von Helmholtz.

Of course, many influential philosophers—from Wilhelm Windelband (1848–1915) to Heinrich Rickert (1863–1936), from Paul Natorp (1854–1924) to Ernst Cassirer (1874–1945), Frege to Husserl, Wilhelm Dilthey (1833–1911) to Vladimir Lenin (1870–1924)—soon disagreed with these relativistic uses of natural-scientific results and attitudes. These philosophers sought to refute "biologism," "naturalism," "materialism," or "psychologism"—all taken as so many species of relativism or even skepticism. At the same time, the critics were careful not to appear anti-scientific; they were objecting to what they regarded as a mistaken over-extending of natural science into the domains of the humanities in general and philosophy in particular. The attacks were often combined with the lament that too many philosophical chairs had already been taken over by scientists posing as philosophers. Frege's and Husserl's arguments against psychologism are today the best-known contributions to this genre. For both men, logical laws were ideal and outside of space and time. Thus, these laws could not be studied with the methods of empirical science. Indeed, all scientific work always already presupposed logical laws.

The argument did not end there. In response to Husserl and other critics of relativistic naturalism, a number of authors proposed non-relativistic ways of overcoming a strict separation of the empirical and ideal domains. Such proposals flourished first and foremost in and around psychology and especially in the Weimar period.

The four papers published in this section selectively highlight four key junctures of the development sketched above in broad outline.

Lydia Patton focuses on how the pioneers of psychophysics, Ernst Weber (1795–1878) and Gustav Fechner (1801–1887), as well as their most important interpreter, von Helmholtz, developed the idea that perception is physiologically, psychologically, and perspectivally relative to the human observer. Patton also shows that this idea had important consequences for how these authors thought of scientific knowledge.

Richard Staley studies the relationship between "physical relativity" and philosophical relativism in Mach and Einstein. He argues that both men's uses of the term "relativism" was influenced not only by debates in physics but also by their reflections on science and politics and views of absolute and relative across disciplines from history to physics.

Dermot Moran discusses Husserl's life-long engagement with relativism from the *Logical Investigations* to the late *Crisis* writings. Relativistic naturalism and historicism were the central targets throughout Husserl's *oeuvre*. Moran's paper is also significant for other sections of this book: e.g., Husserl's attack on Dilthey and historicism is important for the "history" section, and his attack on Lucien Lévy-Bruhl (1857–1939) for the "society" section.

Finally, Paul Ziche discusses various attempts in the period around 1900 of steering a middle path between absolutist idealism and psychological empiricism. The central theme was that philosophy and the natural sciences could live in harmony as long as the latter (and their philosophical interpreters) gave up on naturalistic reductionism.

4 Perspectivalism in the development of scientific observer-relativity[1]

Lydia Patton

Enlightenment perception

The history of early modern empiricism and idealism in philosophy is entwined with normative accounts of rationality and perception.[2] When René Descartes, John Locke, and David Hume are faced with the objection to their theories of perception that not all human beings experience things in the same way, they appeal to the "healthy" adult human, free of "disease," "sane" of mind, and without any perceptual differences such as colorblindness. "Normal" human beings are able to engage in acts of perceiving and reasoning that follow, or become, normative standards. Those whose perceptions do not meet these normative standards are considered to have an overactive imagination, which is associated with mental illness in the writings of Locke, Descartes, and Malebranche.[3]

The "Enlightenment ideal of rationality" may be a myth of the scholars. But it has a basis in the texts, and, as Hatfield (1990) has emphasized, it has a counterpart in normative standards for human perception. To reason competently about the world, it is not sufficient to be able to make correct inferences: one must also be able to *perceive* the phenomena "correctly." In many authors of the seventeenth and eighteenth centuries, the standard of correct perception is relative to the perceptual capacities of human beings.

There is an important limitation of this claim, in both the Cartesian and Lockean traditions. A competent perceiver of Lockean ideas is able to perceive not only the primary qualities of things that are features of the mind-independent world but also those qualities that are secondary: qualities of our perceptions that arise in an interaction between the perceiver's sensory capacities and her environment and that depend constitutively on both. But arguably, for Locke, the perception of primary qualities does not *depend* on the perception of secondary qualities, and so we do not need to give an account of observer-relative qualities when giving an account of how knowledge is obtained through perceptual experience.[4]

The Cartesian knower shuts her eyes to all unclear and indistinct physical perceptions in order to see the true ideas by the natural light of reason— and these ideas are not relative to the perceiver.[5] In this sense, Descartes and

Locke agree: True ideas do not depend on properties of the perceiver or of her sensory capacities.

They disagree about whether sensation can reveal real qualities of things. Descartes denies that veridical perception of sensible qualities, whether primary or secondary, reveals the true qualities of the world as it is. As Ott (2017) reads the text, in *The World, or Treatise on Light*, Descartes argues against the Scholastic view that "the sensible qualities that we experience either just are or resemble the sensible qualities that exist in the world" (1664, §2.3.1). Instead, as Descartes remarks:

> Words, as you well know, bear no resemblance to the things they signify, and yet they make us think of these things, frequently without our even paying attention to the sound of the words or to their syllables ... Now if words, which signify nothing except by human convention, suffice to make us think of things to which they bear no resemblance, then why could nature not also have established some sign which would make us have the sensation of light, even if the sign contained nothing in itself which is similar to this sensation? Is it not thus that nature has established laughter and tears, to make us read joy and sadness on the faces of men?
>
> (1664, 79)

Cartesian sensible qualities do not directly reflect the qualities of mind-independent objects. Instead, they stimulate us to think of *other* things, as when we think about the taste of pie when we see an advertisement for pie. A flat, odorless picture of a pie does not have any resemblance to the taste of lemon meringue, but it is "natural" to make a mental association between one and the other. Similarly, in Descartes' and Locke's accounts, the particles streaming from the pie and impinging on the subject's senses do not resemble the pie. But they stimulate the mind to form an idea of the pie. In Malebranche's elaboration of the Cartesian philosophy, there is divine guidance of the mental process of inference from sensation to idea.

Psychophysics: Qu'est-ce que c'est?

To a physiologist of perception, that is, to someone interested in the experimental and physical basis of the act of perception, Descartes' and Locke's accounts leave parallel questions unanswered. Descartes argues that there is a "natural" relationship between sign and thing signified, but he does not explain—in *The World*, at any rate—how that relationship can be investigated experimentally. Locke argues that the sensible, primary qualities of objects are real qualities of those objects, but he does not provide a physiological explanation of this claim. Physiologists of perception seek an experimentally verifiable, or at least an empirically well-founded, explanation for any relationship between sensed and objective qualities. From the perspective of the physiology of perception, Locke and Descartes appeal to a natural or pre-established harmony between sensation and idea. The difference is in the ground they postulate for the inference from one to the other: Cartesian "natural geometry" or Malebranchian divine natural judgments, or Lockean habitual inference from experience.

From this perspective, Kant's Critical philosophy introduces a twist on the explanation of visual experience, by emphasizing the role of sensibility, perception, and representation in epistemology. In the *Critique of Pure Reason* (1781/1787), Immanuel Kant asks how knowledge can be relative to human sensibility, how our acquisition of knowledge may depend constitutively on purely subjective features of our experience, and yet how knowledge can be universal, necessary, and objective. What Descartes saw as a "natural" harmony between subjective signs and ideas is a problem for Kant: the problem of how to interpret subjective sensations as evidence for inferences that support objective knowledge.[6] These two ideas are not inconsistent. In fact, one could see Kant as responding to the Cartesian problem.[7] How are we to find a logic of perception: a formal account of how nature and reason lead us from sensation to knowledge? Hatfield (1997) proposes that we understand the "the development of philosophy from Descartes to Kant ... as a series of claims about the power of the intellect to know the essences of things, with resulting consequences for ontology and for the role of sensory cognition in natural philosophy." On Hatfield's reading,

> Kant entered his critical period when he realized that human cognizers do not have available the 'real use' of the intellect or understanding to know an intelligible world of substances; at the center of his critical (theoretical) philosophy was his new theory of the human understanding as a faculty limited to synthesizing the materials of sensory representation but unable to penetrate to things in themselves.
>
> (1997, 22)

One can note as well that, in the move from Descartes to Kant, the *idea* is no longer a standard derived from our knowledge of essences. The standard for knowledge in Kant comes via proofs of the validity and objective reality of a set of rules of synthesis of representations, which ground knowledge from experience. Locating the ground of objective knowledge in a relation between the subject and the object, rather than in the subject's knowledge of ideal essences, has a well-known and profound effect on natural philosophy.

The epistemic relation between sensation, perception, inference, and knowledge came to the forefront in physiological neo-Kantianism, and in the closely entwined tradition of empirical psychology, from the beginning to the end of the nineteenth century.[8] Kant's focus on the conditions for objective judgments put an emphasis on how to determine the relationship between subjective and objective in perception and in knowledge.[9]

Much of the research in German psychology done at the time, including that of Johann Friedrich Herbart and Wilhelm Wundt, focused increasingly on what could be investigated empirically. In which circumstances, if any, can we consider a subjective perception to be evidence of the properties of things-in-themselves? If it is not possible to have knowledge of things-in-themselves, as Kant and many neo-Kantians argue, we may decide instead to delineate the contributions of the subjective and the objective, and to show how each operates in phenomenal experience.

An explanation of phenomenal experience along these lines will allow us to identify those aspects of experience that are objective, stable, and manipulable, which is sufficient for a scientific account. But it is not necessarily the case that identifying what is "objective" requires disentangling the objective part of experience from the part that is relative to the observer. To the post-Kantians, knowledge can be "objective," "scientific," and "relative to the observer" simultaneously.

The tradition of "psychophysics" established by Ernst Weber and Gustav Fechner set the stage for mid-nineteenth century work in this area.[10] Psychophysics establishes quantitative relationships between qualitative sensations and their stimuli, and investigates the dynamics of these sensations, including how they arise and recede, and how they are heightened or dulled in response to stimulus. Researchers in psychophysics must rely on results from physiology of perception to establish standards of measurement for sensation, and on results from physics to establish differential equations describing the variation of sensation with respect to stimuli.

The contributions of the tradition of psychophysics go beyond the quantitative analysis described above, however. Influenced by the questions described above, which developed in the tradition of natural philosophy including the work of Descartes, Locke, Leibniz, and Kant, the founders of German physiology of perception and psychophysics explicitly set up their research to answer epistemological questions.[11]

In *The Sense of Touch and the Common Sense*, Ernst Weber was concerned to establish which elements of sensation were objective and which subjective. Gustav Fechner (1859) and Weber established the well-known Weber-Fechner law, which is a quantitative stimulus-response relation. But the epistemological investigations of psychophysicists went well beyond the mathematical establishment of a stimulus-response curve. Fechner's *On a Fundamental Law of Psychophysics and its Relationship to the Estimation of Stellar Size* analyzes the difference between the apparent and the real diameter of stars.[12] According to optical theories, that is, theories of light itself and of our perception of it, we should see the stars as they are. Light emanating from the stars has a certain wavelength. The light strikes our retinas, and the optic nerve transmits the resulting impulses to the brain. As far as this physical system is concerned, there should be no difference between the "phenomenal" and the real diameter of stars. But that is not the case: there are real, measurable differences between the apparent and the real diameter of stars. Nonetheless, these differences are not random. They are stable among the population of human perceivers. Weber begins *The Sense of Touch* with a variant of Fechner's astronomical example:

> We must distinguish, in all sensations, between pure sensation and our interpretation of [sensations]: the sensations of darkness, light, and of colors are pure sensations; that something dark, light, and colored either is in us, or is in space before us, and has a form, is resting, or is moving, is an interpretation of [sensations]. *This interpretation is so closely associated with sensation that it is inseparable from it and we take it for a part of sensation, whereas, in fact, it is a representation that we make for ourselves from sensation.* Not only veridical, but also false interpretations of sensations are mixed in with [sensations], in some cases so closely that one cannot separate [false interpretations] from [sensations], even when one is aware of the error and of the cause of the error. To everyone, even astronomers, the rising and setting sun and the rising and setting moon seem to have a larger diameter than when either of them are high in the sky... the visual angle under which we see these celestial bodies in the two cases is, as measurement proves, exactly the same, but it [the illusion] rests on *a false interpretation that anyone in these circumstances would be forced to make*, so possibly no one yet has been able to be free of it.
>
> (Weber 1905, 4–5, emphasis added)

Mathematical perspective and judgments of distance influence perception, as much earlier theorists including Al-Haytham and Plato recognized. Weber and Fechner take an inferentialist perspective on this phenomenon, arguing that sensations are interpreted to yield perception, and that this perceptual interpretation may be in conflict with independent measurements.

Helmholtz's epistemology

The physiologist of perception and physicist Hermann von Helmholtz was among the first, if not the first, to recognize the novelty and potential of Weber and Fechner's approaches. Helmholtz spent much of his early career investigating perceptual phenomena including stereoscopic vision and the horopter effect. Stereoscopic vision is the phenomenon that humans with two eyes see a single visual image, which is made up of two independent retinal images combined into one. Of course, the retinal images are also upside down, and the brain interprets them as right side up. For Helmholtz, the vast majority of our perceptual experience is an *effect*, caused by the interaction between external objects and our sensory and nervous system.

But this conclusion raises a striking question for epistemology: what is the epistemic status of statements about perceptual experience? Fechner and Weber had argued that many inferences made from sensation are inescapable, even to those who know that they are illusions. In that case, Helmholtz asked, are there regularities in our inferences—even in the incorrect ones—that can be the source of knowledge *about perceptual experience itself*?

There might be three sources of such knowledge, at least:

(1)   A description of the *physiological* facts about perceptions, explaining how sensations arise and interact.

(2)   A *psychological* account of how concepts and inferences may contribute to perceptual experience.

(3)   A *perspectival* theory of how the subject's situation in space, time, and history influences her experience. For instance, being born on a planet with no gravity might influence the inferences from sensation that a subject would make.

Helmholtz provides detailed accounts of the first two of these sources. In the case of the perspectival theory, he does defend the view but does not give a detailed account of how it will work.

The elements (1), (2), and (3) above are combined in Helmholtz's epistemological account of knowledge via perception. For Helmholtz, the spatial and temporal order of sensations is the ground of a "remarkable effect," namely, that objects appear to have sensible qualities—even though we do not perceive those qualities directly. In his lecture "The Facts in Perception," Helmholtz writes:

> Thus, that this intuited spatial order of things originally arises from the sequence in which the qualities of the sensations of the moved sense organ are presented ultimately remains a remarkable effect, even in the accomplished representations of the experienced observer. That is to say, the objects present in space appear to us clothed with the qualities of our sensations. They appear to us red or green, cold or warm, smell or taste etc., while in fact these sensory qualities belong only to our nervous system and certainly do not reach out to external space. Even if we know this, the appearance[13] does not cease, for this appearance in fact is the original truth; it is the sensations themselves, which primarily present themselves to us in spatial order.
>
> (1878, 21)

All qualities are qualities of bodies that are constituted by the properties of our sensory nerves and nervous system.[14] The qualities of perceived objects are the qualities of our sensations, and yet they "clothe" the objects present to us in observation. Helmholtz refers to the fact that we experience perceptual qualities, and not just mechanical sensations, as a *Folge* (effect or consequence).

But this "appearance" is the "original truth." Our perceptual experience is caused by physical and physiological regularities, which includes the stable features of our interaction with our environment.[15] If I perceive an object as red, that quality does not "reach out to external space." We do not have definitive reason to believe that redness is a mind-independent property of the object that caused the sensation. But that does not mean that my perception of redness is an error. It is possible to know what Helmholtz calls *Thatsachen* (facts) about our experience of redness, without ascribing those facts to an object that is independent of the sensing body or mind (Helmholtz 1879). That is the sense of Helmholtz's title *Die Thatsachen in der Wahrnehmung:* the facts *in* perception. Facts can include the claim that "Vermilion is red:" for Helmholtz, this claim can be factual as perceived, but not an "objective" truth in Kant's sense of valid for all perceivers. This will be discussed in more detail below.

Helmholtz defends a second thesis influenced by psychophysics: features of our perceptual experience are constituted by unconscious inferences, which in turn are based on prior experience. Perceptual experience thus is *situated*: perceptual experience can be captured fully only by giving a historical account of previous perceptions, inferences from those perceptions, and their logical and occurrent impingement on present perceptions and representations. Recall the passage from Weber that analyzes the illusion of the setting sun and moon: "not only correct, but also false interpretations of sensations are mixed, in some cases so closely that one cannot separate [the false interpretations] from [the sensations] at all." Even experienced observers—and even astronomers!—experience the setting sun as larger than the sun at high noon, even though they know it is not. This is a result of an inference, but an inference that takes place very swiftly and without our noticing it.

To use another example that Helmholtz gives, I perceive objects that are farther away as smaller than identical objects that are closer to me. This is a mere appearance. If I measure the objects, I will discover they are identical. But it is also an appearance that is forced on me, not just by the nature of my sensory processing system, but by unconscious inferences I am constrained to draw by the nature of my previous experience. The passage from the *Handbook of Physiological Optics* in which Helmholtz describes these inferences is significant to anyone who has been reading Weber and Fechner:

> The mental operations through which we come to the judgment that a particular object in a particular state in a particular place outside us is present, are in general not conscious operations, but unconscious. In their results, they are similar to an *inference*, insofar as we achieve from the observed effect (*Wirkung*) on our senses the representation of a cause of this effect, whereas, in fact, we can only perceive directly the nerve stimulations, that is, the effects, never the external objects. However, they appear to be distinguished from an inference—this word taken in its usual sense—because [an inference] is an act of conscious thought. Such actual conscious inferences are, for instance, when an astronomer calculates the position in outer space, its distance from the Earth, and so on from the

perspectival images that are given to him of the stars at different times and in different positions of the Earth's orbit. The astronomer supports his inferences with a conscious knowledge of the theorems of optics. Such a knowledge of optics is missing in the usual acts of sight. However, it is permissible to describe the mental acts of usual perception as *unconscious inferences* ....

(1867, 430)

A human being moving among the objects she experiences is an earthly astronomer. She eyeballs the measurements of objects in her visual field, and makes comparisons between them, in order not to step into a busy street or to fall off a cliff. This requires complex calculations about spatial relationships, relationships that, according to Helmholtz, are inferred. Those inferences are inferences on the basis of sensations, which are just the stimulations of nerves in a certain sequence, to the size, configuration, and position of external objects.[16] Weber's astronomer still sees the setting sun as bigger than the sun at noon, like the rest of us. Helmholtz's earthly astronomer, the human observer, cannot stop making unconscious inferences in perceptual experience. Even if you *know* that two ships are exactly the same size, you will still *perceive* one that is much farther away as smaller. The inference that grounds this element of our perceptual experience is not a free act, it is an effect of a cause.

In his essay "The Facts in Perception," Helmholtz raises the question: "What is truth in our representations?" This question relates truth or epistemic justification, not to bare sensation but to our representation of external objects on the basis of that sensation. Helmholtz argues that it is possible to explain how we construct representations of objects and processes on the basis of sensations that are experienced in a time order. In the *Handbook of Physiological Optics* and in *On the Sensations of Tone*, Helmholtz develops theories of sound and color. There, Helmholtz gave significant attention to the problem of distinguishing subjective from objective,[17] and to the problem of giving a "physiological" and "psychological" account of the phenomena encountered in conscious experience.

Helmholtz argues that complex qualitative and quantitative features of phenomenal experience, including separation in space, shades of color, and gradations of sound, are not sensed but inferred from sensation. We must make inferences from our sensations, and from the sequence in which they present themselves to us, to have access to a set of qualitative and quantitative features of perceptual experience (Helmholtz 1868, 175–176). We do not perceive the external objects that cause our sensations directly. Rather, we infer from the assumption of a causal interaction between the subject and the object, and from the sensations via our nerve endings, that external objects are present:

The mental operations through which we come to the judgment that a particular object in a particular state in a particular place outside us is present, are in general not conscious operations, but unconscious. In their results, they are similar to an *inference*, insofar as we achieve from the observed effect on our senses the representation of a cause of this effect, whereas, in fact, we can only perceive directly the nerve stimulations, that is, the effects, never the external objects.

(Helmholtz 1867, 430)

Helmholtz's account blocks the appeal to direct "confirmation" or "verification" of beliefs about objects via observation. Perception of external objects is always mediated by inference.[18] Thus, as Hatfield notes, Helmholtz rejects the direct scholastic and Lockean inference from perceived qualities to real qualities of objects. Helmholtz argues that we must *discover* the relationship between sign and object by investigating the stable relationships between sequences of perceptions and changes in the stimuli that produce them, whether these changes are artificially manipulated for experiment's sake, or natural.

Those relationships must be discovered and analyzed within experience itself. Helmholtz does not allow for a standpoint outside experience that serves as a standard. The subject's experience is experience *of* objects, and also reflects the subject's physiology, psychology, and perspective. It is not the case that we must restrict ourselves to introspection when studying experience: all the scientific tools of investigation can be brought to bear, including experiments on how subjects perceive and stimuli to which they respond. By analyzing experience in this way, we can come to have increasingly better knowledge of what is subjective, and what is objective, relative to a particular subject's perceptual experience. "Subjective" and "objective," in this case, are not absolute categories as they are in Kant's sense: there is no "pure" subjectivity in Helmholtz. Rather, Helmholtz's "subjective" and "objective" are explanatory categories, used to construct explanations of processes and elements in perceptual experience that can be ascribed to the subject or to the object. When constructing these explanations, one *a priori* assumption is necessary: that objects outside us exist and that they cause our sensations of them.[19]

A scientific analysis of perceptual experience as observer-relative

There are three senses in which perceptual experience, as described by German empirical physiology and psychophysics, is relative to an observer. One is *physiological*: human perceptual experience depends on human sensory capacities. Another is *psychological*: since habitual judgment impinges on experience, human perceptual experience depends on, and must be analyzed relative to, human psychological and inferential capacities. Yet another is *perspectival*: Again, since habitual judgment impinges on experience, the historical, environmental, and physical conditions for the formation of a subject's *habitual inferences* have a constitutive influence on perceptual experience.

The traditions of psychophysics and of the physiology of perception were entangled over the nineteenth century with long-standing problems from

Cartesian and Lockean natural philosophy, and from the physiological and epistemological strands of neo-Kantianism. The history of psychophysics illuminates possible approaches to these questions. According to the methods used by this tradition, observation is a natural interaction with, and adaptation to, the environment, which should be analyzed as a process. Knowledge gained from experience is always relative to the human observer.

For Helmholtz, that knowledge gained through experience is relative to the observer is precisely *why* the observer-knowledge relation can be analyzed scientifically. The natural process by which humans obtain knowledge is itself subject to a rigorous, experimental analysis, the scientific basis of which grew stronger over the nineteenth century.[20]

The availability of facts about the physiology, psychology, and perspectival basis of perceptual experience is the ground for claims of knowledge from perception that is not independent of the context in which it arises. Such claims are the basis of what is now known as "contextualist knowledge," according to which "A sentence is true for X if and only if it is true as uttered by X, true relative to a context in which X is the speaker" (Williamson 2005, 92). Propositions about sensory and perceptual experience are contextually true for Helmholtz. For instance, the proposition "The mountains look far away" is true for someone who is far away from the mountains, and false for someone close to them. But Helmholtz's position goes deeper than this. He even argues that the statement "Vermilion looks red" is not true in an absolute sense, but only relative to our sensory faculties—even to the faculties of a particular observer:

> To ask whether vermilion is actually red, as we see it, or whether this is a sensory illusion, is … senseless. The sensation of red is the normal reaction of normally formed eyes to the light reflected by vermilion. A color-blind person would see vermilion as black or dark grey-yellow; this too is the correct reaction for a different eye … In itself the one sensation is not more correct or more false than the other.
>
> (Helmholtz 1867, 445)

This is a classically contextualist position. "This vermilion looks red" is true for observer A, who is not colorblind, but not for observer B, who *is* colorblind.

The contextualist reading of Helmholtz, strongly supported by the passage above and by many of Helmholtz's remarks, raises the question: If Helmholtz is a contextualist, what does this mean for Helmholtz's epistemological account of knowledge from perception? We might try to situate Helmholtz's view within contemporary positions on color realism, for instance. Here, I would advise caution. Among the positions taken in the contemporary context are realism (Byrne and Hilbert 2017), relationalism (Cohen 2009), and relativism. Relationalism is the view that, for certain properties like being poisonous or having a certain color, "There is no such thing as [having that property] simpliciter. Rather, there is a family of relational properties," defined in relation to a subject: something can be poisonous to humans but not to snakes, and something can appear green to me but not to you (Byrne and Hilbert 2017, 173). Relationalism, as Byrne and Hilbert observe, "multiplies perceptible properties" (2017, 175). There are a number of perceptible properties in the world: green-to-Lydia is a distinct property from green-to-Hermann, for instance.

We might, then, try to locate Helmholtz within the spectrum of "relationalist" theories of color, as against realist theories, according to which there is a property, "redness," that exists independently of human observers. Tracz (2018) is a cogent defense of the relationalist reading. While the relationalist reading of Helmholtz is certainly defensible, it is not the end of the story: restricting ourselves to an argument according to which Helmholtz *only* is defending contemporary relationalism would risk losing some of the force and complexity of Helmholtz's view.

Tracz (2018) makes a convincing case that Helmholtz defends a view close to Cohen's relationalism about perceptual properties. The relationalist view is intended to answer questions about "the metaphysical status of the properties that we perceive" (Tracz 2018, 66). Relationalism is the view that there are families of relational properties, as Byrne and Hilbert put it, in terms of which we can explain the metaphysical status of perceptual properties.

Contemporary relationalism seeks to define a set of *stable* relational properties. While it is true that one can become blind, or one's sense of smell and thus taste can change, it is also the case that, for Helmholtz, *habit* and *inference*—and even deliberate interference—can change the relational properties involved, and thus change my perceptual experience. For Helmholtz, humans are detectors of a certain kind: human ears can perform Fourier analysis, and human eyes detect radiation at specific wavelengths.[21] But for Helmholtz, Fechner, and Weber, perceptual experience of complex phenomena like music and colored objects is not reducible to the ground-level sensory response to physical stimuli. Inference is inextricably involved in human perceptual experience. Helmholtz took from psychophysics the idea that my previous experience impinges on my perceptual experience, and Helmholtz argues that the impingement happens through a series of inferences that can be manipulated experimentally.

Certain judgments and perceptions are stable features of our perceptual experience. We perceive objects that are farther away as smaller than identical objects closer to us. We perceive the setting sun as larger than the sun at noon, but it is the same sun. We reliably can produce stable perceptual illusions by manipulating sensory and motor responses to stimuli (Helmholtz 1867, 429, 436–437).

More recent research supports the idea from Helmholtz, Weber, and Fechner that some sources of the processing that results in perceptual experience arise from changing experience, not just from stable—or even relatively stable—properties of a perceiving subject. The psychologist Diana Deutsch details the activity of tonal processing (1999), including illusions that arise in the recognition of music (1969) and auditory illusions in general (1974, 1975). Auditory illusions arise that can be associated with right- and left-handedness: right-handed persons perceive certain pitch combinations differently from left-handed persons. Deutsch's research supports the view from Helmholtz, Weber, and Fechner that experience, in conjunction with physiological and psychological factors, can influence perceptual processing.

As I have argued throughout this essay, Helmholtz allows for not only physiological facts and psychological inferences but also perspectival reasoning, to influence perceptual experience and knowledge gained from perception. But Helmholtz also defends a version of the view according to which there can be a kind of "perspectival truth" revealed in scientific research and investigation. Helmholtz argues that the relationships between subjective and objective, real and actual, actual and illusory, must be analyzed scientifically, within experience. There is no standpoint outside experience from which we can reason, no extra-sensory knowledge of the constitution of the "ideal subject" or of the properties of "real objects." Instead, we reason using the "sign system" of our sense impressions, which allow us to form representations of objects. Those representations are always physiologically, psychologically, and perspectivally relative to the observer. However, precisely for that reason, they can be used as scientific evidence, provided that relativity to the observer *itself* can be analyzed scientifically. In the tradition of psychophysics inherited by Helmholtz, we can arrive at a kind of perspectival analysis of perceptual experience, which embeds an account of that experience within the context of the history and situation of the perceiving subject.[22] That analysis is relative to the perceiving subject, but the perspectival explanations Helmholtz constructs are not thereby empty of scientific significance: in fact, for Helmholtz, the more squarely the perceiving subject is placed in a scientific, perspectival context, the more facts we are able to learn about her experience and the objects with which she interacts.

Notes

1 I would like to thank Martin Kusch, Katherina Kinzel, Johannes Steizinger, and Niels Wildschut for the opportunity to contribute to this volume, and for substantive comments on an earlier draft, which have influenced the project of the chapter much for the better. Walter Ott has provided valuable contributions to the chapter's account of Locke, although, of course, he is not responsible for the details of the interpretation that I have given here.
2 Hatfield (1990) is a classic reference in this context.
3 These remarks are familiar to those who study this period. They occur in Descartes' *Meditations* 4 and 6, in Malebranche's *The Search after Truth* (for instance, 2.3.1.4, 3.1, and 3.2), and in Locke's *Essay Concerning Human Understanding* (for instance, 2.11.13). See Hume (1739), especially the editorial material from the Nortons on p. 751 and p. 771.

4 As Walter Ott conveyed in personal communication, Locke's "simple" account of the perception of primary qualities can be found in the *Essay* (Locke 1689), especially II.v and II.ii.1. The first eight chapters of Book II of that work present Locke's theory of visual experience. As Ott notes, chapter nine complicates the picture by adding judgment: "Because Sight, the most comprehensive of all our Senses, conveying to our minds the *Ideas* of Light and Colours, which are peculiar only to that Sense; and also the far different *Ideas* of Space, Figure, and Motion, the several varieties whereof change the appearances of its proper Object, *viz.* Light and Colours, we bring ourselves by use, to judge of the one by the other. This in many cases, by a settled habit, in things whereof we have frequent experience, is performed so constantly, and so quick, that we take that for the Perception of our Sensation, which is an Idea formed by our Judgment; so that one, viz. that of Sensation, serves only to excite the other, and is scarce taken notice of it self; as a Man who reads or hears with attention and understanding, takes little notice of the Characters or Sounds, but of the *Ideas,* that are excited in him by them" (Locke 1689, II.ix.9, 146–147). It may seem *prima facie* as if this passage supports a reading of Locke that is closer to Descartes: that our perceptions are a language we must read, not direct evidence of the properties of bodies. However, on Ott's reading, Locke introduces judgment into his account of visual experience to correct for mistakes in the original experience that cloud our understanding of that experience. And the passage above bears this out: the rhetorical force of the passage is to argue that when we begin to habitually associate ideas of light and color with ideas of space, figure, and motion, ideas from one place (judgment) are being *wrongly* used to judge ideas from another (sensation). Unlike Descartes, Locke believes that an analysis of sensation can clear up any confusion between judgment and sensation and can reveal the evidence of primary qualities that is given in perceptual experience. On a different subject, several ideas resembling those in this passage are discussed by Helmholtz in §26 of his *Handbook*.

5 In *Madness and Civilization*, Foucault writes memorably: "the Cartesian formula of doubt is certainly the great exorcism of madness. Descartes closes his eyes and plugs up his ears the better to see the true brightness of essential daylight; thus he is secured against the dazzlement of the madman who, opening his eyes, sees only night, and not seeing at all, believes he sees when he imagines. In the uniform lucidity of his closed senses, Descartes has broken with all possible fascination, and if he sees, he is certain of seeing that which he sees" (1961, 102).

6 The reading of Kant here owes much to Hermann Cohen and the Marburg School. For an early version of my reading of the Marburg School, see Patton (2004).

7 To be sure, Descartes was far from the first to pose this problem. Plato's *Timaeus* contains a beautiful statement of a similar question. In classical Indian philosophy, we find a fascinating debate between Buddhist and Hindu thinkers of the Vasubandhu and Nyāya traditions, on the question of how perception can become veridical cognition, and on the status of perception and sensation themselves (see Chadha, 2016).

8 For more on the tradition of "physiological neo-Kantianism," see Patton (2004) and Beiser (2014), including references there to others' work. As Hatfield details (2018, §4.1), the question of the relation between the subjective and the objective in perception arose even for empirical physiologists of perception, like Johann Georg Steinbuch and Caspar Theobald Tourtual.

9 See Edgar (2013, 2015).

10 Fechner's psychophysics owes a debt to the earlier work of Johann Friedrich Herbart. Michael Heidelberger has written the definitive work on Fechner with *Die innere Seite der Natur* (*The inner side of nature*). In 1863, Ernst Mach delivered lectures on psychophysics, which are published in German; an appreciation of the lectures in English is in Titchener (1922).

11 While recent interpreters of Helmholtz's work disagree on the influence of Fichte on Helmholtz, they appear to agree on *this* point: that Helmholtz was concerned to use the methods of empirical physiology to answer epistemological questions, and even to dissolve metaphysical questions that turn out to be empirical questions in disguise. Compare, for instance, de Kock (2015) and Heidelberger (2015) to Hatfield (2018).

12 Original title: *Über ein psychophysisches Grundgesetz und dessen Beziehung zur Schätzung der Sterngrössen* (1859).

13 The word *Schein* has been translated as "illusion" in the past, but I now believe this translation to be misleading.

14 Helmholtz discusses the example of the setting moon in §30.

15 This includes features of our voluntary motions with respect to objects, as de Kock emphasizes (2014, 2015).

16 "Now, the described unconscious inferences from sensation to their causes are congruent in their results to the so-called *inferences from analogy*. Because in a millionfold majority of cases, stimulation of places on the retina on the outer corner of the eye originates from external light that falls on the eye from the area of the bridge of the nose, we judge that it will be so as well in each newly encountered case in which the stated places on the retina are stimulated, just as we assert that each single human now living will die, since experience has revealed that all humans living in the past are dead. Further, these unconscious inferences from analogy arise with compulsory necessity, since they are not acts of free conscious thought, and their effect cannot be reversed through better insight into the connection of things" (Helmholtz 1867, 430).

17 Emphasized by Ernst Weber in *The Sense of Touch* (1834) and in *The Sense of Touch and the Common Sense* (1846).

18 "We use the sensations that light stimulates in our apparatus of sensory nerves, to form for ourselves representations from them [the sensations] concerning the existence, the form, and the location of external objects. We call such representations *visual perceptions*. … Since perceptions of external objects thus belong to the representations, and representations always are acts of our mental operation, perceptions can come about only in virtue of mental operation, and thus the doctrine of perceptions in fact already belongs to the domain of psychology" (Helmholtz, 1867, 427).

19 See Patton (2009) for a discussion of Helmholtz's reasoning on this score.

20 For accounts of the development of research in nineteenth-century physiology of perception in the labs, see Finkelstein (2013), Otis (2007), and Sulloway (1992).

21 Lenoir (2006) illuminates Helmholtz's experimental practices in testing these views, building material objects like Helmholtz resonators to stand in for human sensory apparatus.

22 The most recent and compelling argument for "perspectival truth" is presented by Michela Massimi (2018). Massimi argues for perspectival realism as a version of scientific realism. It is possible that the work of Grete Hermann provides an early version of this (see Banks 2017). See also Brogaard (2010) for an argument in the case of color.

References

Banks, E. (2018), "Grete Hermann as Neo-Kantian Philosopher of Space and Time Representation," *Journal of the History of Analytical Philosophy* 6 (3): 244–263.

Beiser, F. (2014), *The Genesis of Neo-Kantianism, 1796–1880*, Oxford: Oxford University Press.

Brogaard, B. (2010), "Perspectival Truth and Color Primitivism," in *New Waves in Truth*, edited by C. Wright and N. Pedersen, Basingstoke: Palgrave-Macmillan, 1–34.

Byrne, A. and Hilbert, D. (2017), "Color Relationalism and Relativism," *Topics in Cognitive Science* 9: 172–192.

Chadha, M. (2016), "Perceptual Experience and Concepts in Classical Indian Philosophy," *The Stanford Encyclopedia of Philosophy*, edited by E. N. Zalta, URL = <https://plato.stanford.edu/archives/spr2016/entries/perception-india/>.

Cohen, J. (2009), *The Red and the Real*, Oxford: Oxford University Press.

De Kock, L. (2014), "Voluntarism in Early Psychology: the case of Hermann von Helmholtz," *History of Psychology* 17 (2): 105–128.

—— (2015), "Hermann von Helmholtz's Empirico-Transcendentalism Reconsidered," *Science in Context* 27 (4): 709–744.

—— (2018), "Historicizing Hermann von Helmholtz's Psychology of Differentiation," *Journal for the History of Analytical Philosophy* 6 (3): 42–62.

Descartes, R. (1664), "The World, or Treatise on Light," in *The Philosophical Writings of Descartes, vol. I*, translated and edited by J. Cottingham, R. Stoothoff, and D. Murdoch, Cambridge: Cambridge University Press, 1985, 79–98.

Deutsch, D. (1969), "Music Recognition," *Psychological Review* 76 (3): 300.

—— (1974), "An Auditory Illusion," *The Journal of the Acoustical Society of America* 55 (S1): 18–19.

—— (1975), "Musical Illusions," *Scientific American* 233 (4): 92–104.

—— (1999), "The Processing of Pitch Combinations," *The Psychology of Music* 2: 349–412.

Edgar, S. (2013), "The Limits of Experience and Explanation," *British Journal for the History of Philosophy* 21 (1): 100–121.

—— (2015), "The Physiology of the Sense Organs and Early Neo-Kantian Conceptions of Objectivity," in *Objectivity in Science: Approaches to Historical Epistemology*, edited by F. Padovani, A. Richardson and J. Tsou, Dordrecht: Springer, 101–122.

Fechner, G. (1859), *Über ein psychophysisches Grundgesetz und dessen Beziehung zur Schätzung der Sterngrössen*, Leipzig: S. Hirzel.

Finkelstein, G. (2013), *Emil du Bois-Reymond: Neuroscience, Self, and Society in Nineteenth-Century Germany*, Cambridge, MA: The MIT Press.

Foucault, M. (1961), *Madness and Civilization*, abridged translation by R. Howard, London: Routledge, 2001.

Hatfield, G. (1990), *The Natural and the Normative*, Cambridge, MA: MIT.

—— (1997), "The Workings of the Intellect," in *Logic and the Workings of the Mind*, edited by P. Easton, Atascadero, CA: Ridgeview Publishing Co., 21–45.

—— (2018), "Helmholtz and Philosophy," *Journal for the History of Analytical Philosophy* 6 (3): 10–41.

Heidelberger, M. (2015), "Naturalisierung des Transzendentalen in der Sinnesphysiologie von Hermann von Helmholtz," *Scientia Poetica* 19 (1): 205–233.

Helmholtz, H. (1855), *Über das Sehen des Menschen*, Leipzig: Leopold Voss.

—— (1867), *Handbuch der physiologischen Optik*, Leipzig: Leopold Voss.

—— (1868), "The Recent Progress of the Theory of Vision," in *Science and Culture*, edited by D. Cahan, Chicago: University of Chicago Press, 1995, 127–203.

—— (1879), *Die Thatsachen in der Wahrnehmung*, Berlin: Hirschwald.

Hume, D. (1739), *A Treatise of Human Nature. Volume 2,* edited by D. F. Norton and M. Norton (eds.), Oxford: Oxford University Press, 2007.

Lenoir, T. (2006), "Operationalizing Kant," in *The Kantian Legacy in Nineteenth-Century Science*, edited by M. Friedman and A. Nordmann, Cambridge, MA: The MIT Press, 141–210.

Locke, J. (1689), *An Essay Concerning Human Understanding*, edited by P. H. Nidditch, Oxford: Clarendon, 1975.

Massimi, M. (2018), "Four Kinds of Perspectival Truth," *Philosophy and Phenomenological Research* 96 (2): 342–359.

Otis, L. (2007), *Müller's Lab*, Oxford: Oxford University Press.

Ott, W. (2017), *Descartes, Malebranche, and the Crisis of Perception*, Oxford: Oxford University Press.

Patton, L. (2004), *Hermann Cohen's History and Philosophy of Science*, Ph.D. dissertation, McGill University.

—— (2009) "Signs, Toy Models, and the A Priori: from Helmholtz to Wittgenstein," *Studies in the History and Philosophy of Science* 40 (3): 281–289.

—— (2014), "Hermann von Helmholtz," *The Stanford Encyclopedia of Philosophy*, edited by E. N. Zalta, URL = <http://plato.stanford.edu/archives/fall2014/entries/hermann-helmholtz/>.

—— (2018), "Helmholtz's Physiological Psychology," in *Philosophy of Mind in the Nineteenth Century*, edited by S. Lapointe, New York: Routledge.

Sulloway, F. (1992), *Freud, Biologist of the Mind*, Cambridge, MA: Harvard University Press.

Titchener, E. B. (1922), "Mach's 'Lectures on Psychophysics'," *The American Journal of Psychology* 33 (2): 213–222.

Tracz, R. (2018), "Helmholtz on Perceptual Properties," *Journal for the History of Analytical Philosophy* 6 (3): 63–78.

Turner, R. S. (1994), *In the Eye's Mind*, Princeton: Princeton University Press.

Weber, E. (1905), *Der Tastsinn und das Gemeingefühl*, second edition, edited by E. Hering, Leipzig: Wilhelm Engelmann.

Williamson, T. (2005), "Knowledge, Context, and the Agent's Point of View," in *Contextualism in Philosophy*, edited by G. Preyer and P. Peter, Oxford: Oxford University Press, 91–114.

5 Physical or philosophical?

Mach and Einstein on being a relativist

Richard Staley

Introduction

> Critic: People like me have often raised concerns of the most varied kind against relativity in the newspapers; but only seldom has one of you relativists answered them.
>
> (Einstein 1918a, 697)

This is how Albert Einstein began his 1918 article "Dialogue on Objections to the Theory of Relativity" in the pages of the general-science journal *Die Naturwissenschaften*, immediately setting the exchange in the sphere of public debate and newsprint but also adding a footnote to state explicitly that "Relativist" denoted an adherent of the physical theory of relativity, not philosophical relativism. Recognizing a strong distinction between the two has often seemed to be the first thing to emphasize about the relations between physical and philosophical relativity, noting that Einstein also flirted with the name absolute or covariance theory (with the mathematician Felix Klein suggesting *Invariantentheorie* in 1910), and that this might have obviated much public confusion (Miller 1981, 175; Hentschel 1990, 92–105). In this chapter, I will develop a more complex perspective by taking up what it meant to be a "relativist" for each of the two figures most responsible for the emergence of physical relativity, the Austrian physicist Ernst Mach and Einstein, showing that both used this term rather briefly and at critical points in the emergence of discussions about the theory among their colleagues. But to understand why the distinction was important—why physical and philosophical relativism might be close enough to be considered at the same time, even as Einstein distinguished them—it will be helpful to consider a still earlier discussion of absolute and relative that crossed the domains of scientific knowledge and historical development with strong political implications.

Writing from London in 1878, Friedrich Engels drew together a series of articles eviscerating a theory of socialism that the Berlin academic and writer Eugen Dühring had based on a general philosophical system. Engels thought nature provided historical and materialistic proof of a dialectic of change that stood in stark contrast to the kind of claims for absolute truth, reason, and justice offered by Dühring and others. Moving among the exact, organic, and historical sciences, Engels engaged a highly interesting tension between relative and absolute knowledge. Given the detailed discussion of his work elsewhere in this volume, I can draw out several critical points by focusing on Engels's treatments of physics. Discussing space and time, Engels touched briefly on the consequences of Galileo's principle of relativity to note that "all rest, all equilibrium, is only relative, only has meaning in relation to one or another definite form of motion" (1878, 40). Thus, a body on earth may be in mechanical equilibrium while also participating in the motion of the earth and that of the solar system and being subject in its microscopic elements to the vibrations of the mechanical theory of heat. Amidst a welter of thoughts on heat, joined later by discussions of geology, Darwinian theory, and more, it would be a mistake to see a physical theory of relativity in these brief references. But they went together with an insistence that there was little room for ultimate or final truths in the physical or life sciences, and even less justice for such claims in the realm where they were most often made, the historical sciences; there, Engels argued strongly, knowledge is "essentially relative" in being limited to the interconnections and implications of social and state forms that are by their nature transitory (1878, 66, 68). Physicists recognized that Boyle's Law holds good only within definite limits of pressure and temperature, and for certain gases; Engels would make a similar point about good and evil, relating views on morality to historical periods and social and economic conditions in different classes (1878, 70, 73).

This brief sketch establishes a number of important points that Christopher Herbert argues at length in *Victorian Relativity*, his study of relativism in Victorian Britain (2001). Herbert shows that forms of relativistic arguments had been fielded widely by John Stuart Mill, Alexander Bain, W. K. Clifford, and others. Often expressing elements of a radical, anti-authoritarian politics and receiving urgent denial from those arguing for absolutes, they were regularly immersed in forms of rhetorical and actual violence. Without closely addressing the discipline of physics, or considering Engels, Herbert argues that these discussions set an important context for the development of relativistic thinking in physics and the reception of Einstein's work—which he discusses in some detail, studying the anti-Semitic and political responses of Philipp Lenard, Johannes Stark and the Nazi party in the 1920s and 1930s (Herbert 2001, Introduction). My reading draws attention to the tension between relative and absolute that Engels described across different fields of knowledge, from the exact natural to the contingent historical sciences. Perhaps historians have neglected these aspects of his thought because Engels was so interested in relating the superstructure of ideology to specific political and economic issues, and especially class. Despite this emphasis, his arguments can well be understood as a variant of philosophical relativity. I will not here consider the reception of Engels's work in any great detail, but examining the framework within which Mach and Einstein developed their approaches to a physical theory of relativity will allow us to establish the nature of its pertinence in two important cases, illustrating the extent to which the broader cultural and political framework that Herbert considers was also significant for their work.

Ernst Mach's relational physics

Soon after finishing his doctorate on electrical induction at the University of Vienna in 1860, Ernst Mach published a series of papers providing empirical and theoretical support for Christian Doppler's argument that perceived tone or color depends on the relative motion of the source of light or sound and the observer (Mach 1860; 1861; 1862; Blackmore 1972, 17–19; Schuster 2005). At the same time, Mach began a second line of research developing further the investigation of the relations between physical stimulus and psychical reaction that Gustav Theodor Fechner had christened "psychophysics" (Fechner 1860; Heidelberger 2004; Ziche 2008). Reviewing a series of studies that showed that just perceptible differences in light intensity were proportional to the stimulus in lectures on psychophysics, Mach commented:

> All these experiments teach us that not absolute, but relative differences are important to us. A stimulus must be increased by a certain aliquot part for the increase to be noticed. Conversely, the simplest observation teaches that one and the same absolute difference in stimulus, depending on the size of the stimulus, may sometimes be noticeable, sometimes unnoticeable. The absolute difference in light between the points of the stars and the surrounding celestial ground is the same in day and at night. But in the daytime intense sunlight, reflected from all parts of the sky, prevents us from noticing the difference.
>
> (1863, 243)

Perhaps the most direct fruit of Mach's general perspective emerged in a combination of pointed arguments that current conceptions of the mechanical laws and absolute time and space were inadequately defined scientifically (despite the fruitfulness of Newton's conceptions) and lacked empirical footing. The solution was to develop relational perspectives. Mach penned a paper on mass, but running up against "the physics of the schools" (Mach 1872, 80) had trouble publishing it in *Annalen der Physik*. He could send it elsewhere but only set out his fuller perspectives in the endnotes to a lecture on the history and roots of the conservation of work published as a booklet in 1872 (Mach 1868, 355; 1872, 75–76, 80). This reflects simultaneously the radical breadth of Mach's critique, his awkward disciplinary situation, and the social and political tensions he was ready to engage.

Mach's lecture was delivered to the *Royal Bohemian Society of Sciences* in Prague in winter 1871. Noting that as a child he had never understood why the world should permit itself to be ruled by a king "for even a minute," or why there should be such inequalities of wealth, Mach stated that you could either grow so accustomed to such puzzles as to forget them, or learn to understand them historically in order to be able to consider them without hatred (1872, 15). His history of the conservation of work argued that results often taken to have been established only with the articulation of the conservation of energy by Helmholtz and others had in fact been assumed in earlier work when Stevin, Galileo, and Huygens relied more or less explicitly on the impossibility of perpetual motion. Recognizing that the argument preceded the mechanical theory of heat and its logical root was more general in being another form of the law of causality, should prevent current physicists from overvaluing mechanics as a foundation for all knowledge (1872, 69–71). A second corollary lay in the details. Discussing equivocations between concepts of substance and motion, Mach traced differences in conceptions of heat and electricity to their paths of development (1872, 44–47). Recognizing that they were contingent, accidents of history, should also enable them to be surmounted. Working across the customary disciplinary methods of the natural and historical sciences to provide a history of physics, Mach sought the means to change physics. After raising questions about the crown and wealth, before embarking on the conservation of work, his introduction entreated the audience: "Let us not let go of the guiding hand of history. History has made all, history can change all." (Mach 1872, 18).

Mach raised his most specific new question about current conceptions of mechanical laws in an endnote to his discussion of Galileo's and Huygens's concepts of inertia. Like Carl Neumann, Mach had noticed that Newton's first law of motion is undefined if a frame of reference or concrete body is not specified in relation to which a body remains at rest or in uniform motion. Neumann agreed with Newton that all motion is absolute and postulated a hypothetical body α to resolve this difficulty in definition. In contrast, Mach pointed to the divergent consequences that flowed from two cases of equivalent relative rotation, in which the earth is regarded as turning on its axis, or the heavenly bodies are considered to spin around it while the earth remains at rest. Although equivalent geometrically, in current conceptions inertial effects like the bulging of the earth at the equator followed from the first conception but not from the second. Instead, Mach argued, the law of inertia must be so conceived that the same thing results in each case, which meant that it must be enunciated so as to refer to the masses of the rest of the universe. Discussing the question at some length, he admitted he had not yet found a fully satisfactory way of achieving this (Mach 1872, 75–80).

In 1872, Mach argued sharply for the need to relate the laws of motion consistently to relative motion while treating current science historically in order to recognize its contingency, after referring pungently to the difficulties of the political absolute in royal rule and economic inequalities. His lecture clearly carries many of the rhetorical resonances that Herbert describes and is consistent with Engels's later historical and more determinedly class-based framing of the relations between the natural sciences and ethics. But readers would be unlikely to see much more than an isolated, somewhat speculative argument for physical relativity in his published lecture.

Eleven years later, Mach gave his discussions of current theory and the law of inertia a different cast in his lengthy study, *The Science of Mechanics: A Critical and Historical Account of its Development* (1883, v–vi). The book's aim was anti-metaphysical, and in a central section Mach first outlined Newton's reasons for distinguishing between true, absolute, and mathematical time and space, and relative time and space—and then declared that as the preceding pages had demonstrated, all mechanical knowledge rested on relative measures and relative motion:

> No one can assert anything of absolute space and absolute motion, these are mere figments of thought which cannot be demonstrated in experience. All our fundamental theorems of mechanics, as we have shown in detail, are experimental knowledge concerning the relative positions and motions of bodies.
>
> (1883, 213)

Now Mach dealt directly with Newton's famous argument to prove the existence of absolute space through the rise of water in a rotating bucket, even though the water is at rest in relation to the bucket's walls (1883, 207–221 on 216–217). Mach argued that Newton had neglected the possibility of explaining the inertial effects displayed by the water through the mass of the earth and the distant stars.

Despite the new stage that Mach had given his critiques, he still thought himself isolated. He was grateful for the support offered by Gustav Kirchhoff's 1876 *Vorlesungen über mathematische Physik: Mechanik* (*Lectures on Mathematical Physics: Mechanics*) with its argument that science aimed merely for an adequate description, and in the event Mach's book found unexpected favor (Mach 1942, 325–326; Mach 1889, viii). A large edition sold out within five years, and in 1886 Mach returned to his concern with the relations among psychology, physics, and physiology in a short and challenging book, *Contributions to the Analysis of the Sensations*, in which a footnote announced that the Ego, the absolute soul, was unsavable (1886, 18). A second edition of *The Science of Mechanics* followed in 1889, and a sixth by 1908. Mach's additions and further commentary show that his views of absolute time and space and inertia attracted attention and won increasing acceptance (Mach 1889, 481–485; 1897, 236–237, ix–x, 253–256). By 1901, he noted his pleasure in finding a staunch ally in J. B. Stallo's 1881 book *The Concepts and Theories of Modern Physics*, and a kindred thinker in W. K. Clifford (1901, xi). These references are important in part because, as Herbert shows, Stallo had offered a particularly wide-ranging account of the significance of what he called a two-fold relativity of qualitative and quantitative determinations for cognition (2001, 89–104). In a book whose 1901 German translation carried a foreword from Mach, Stallo wrote:

> There is no absolute material quality, no absolute material substance, no absolute physical unit, no absolutely simple physical entity, no absolute

physical constant, no absolute standard, either of quantity or quality, no
absolute motion, no absolute rest, no absolute time, no absolute space.

(Stallo 1881, 201; Herbert 2001, 93)

Like Clifford, Stallo set relational approaches strongly against monolithic
dogmatisms and theologically inflected absolutes in ringing terms. His expan-
sive and agonistic imagery may be one reason that in 1908 Mach was ready
to write not just of those who agreed or disagreed with his views about
relative motion, but of "relativists." Mach offered an international list of
eight decided relativists that began with Stallo. While perhaps focused on
mathematicians and physicists who had addressed absolute motion in particular, such as Lange, James Thomson, A. E. H. Love, MacGregor, and Paul
Mansion, Mach's list also included figures with broader epistemological aims,
such as Stallo, Karl Pearson, and Hans Kleinpeter, and he thought it was
surely already incomplete (1908, 257).

Mach's confidence was well placed, but it might seem ironic that he was
unaware that his attack on absolutes had also been taken up in electron theory
and the electrodynamics of moving bodies, in particular by Henri Poincaré
commenting on the work of H.A. Lorentz, and then Einstein. By 1908, this
had led to another small group of researchers who identified themselves as
adherents of the principle of relativity, or of the Lorentz-Einstein theory of
relativity. Mach learned of this new and rather open sense of physical relativity through Hermann Minkowski's September 1908 lecture on space and
time. Asking Philipp Frank to tutor him, Mach soon indicated that his earlier
work was consistent with the direction being taken by Lorentz, Einstein, and
Minkowski (Mach 1872, 94). Famously, Mach supposedly repudiated relativity in the posthumously published preface to an unfinished volume on
optics in 1921, but as Wolters has argued persuasively, this is more likely to be
the work of his son Ludwig, who was influenced by Hugo Dingler and sought
the support of Lenard in a further example of the troubling associations
between anti-Semitism and "German" empiricism (Wolters 2012, 48–49).

Despite the promise these forms of relativity held within physics, in the
course of 1908 and 1909 Mach's views also faced two new attacks that may
have been a price of success. The first and most significant challenge—at least
within physics—came from Max Planck, who had earlier written to Einstein
about the need for the small band who accepted relativity to agree with each
other (Plank to Einstein, 6 July, 1907, Einstein 1993, 49–50). On 9 December,
1908, Planck gave an address to students at the University of Leiden that
began and ended with a challenge to Mach, clearly addressing *Contributions
to the Analysis of the Sensations* as much as any other aspect of Mach's
work. Combatting Mach's views that all knowledge depended on sensation,
the boundaries between the physical and psychological were solely practical
and conventional, and that knowledge sought an economical concordance
between thought and experience, Planck argued the history of science was
one of continual de-anthropomorphization, driven by the unity of theoretical

principles and faith in the reality of the world picture. He finished with the biblical injunction that one could recognize a false prophet by the fruits of their views (Plank 1909). Planck had attacked Mach's earlier anti-atomism but initially referred only indirectly to relativity—though his critiques surely aimed also to separate that theory from the Machian epistemology that had done so much to shape it. Mach's reply in *Physikalische Zeitschrift* showed that he thought Planck had misunderstood many aspects of his views—but also argued simply that if physics needed a church and required a belief in atoms, he did not need to be considered a proper physicist, freedom of thought was more important (1910); Wegener has offered a persuasive account of the significant grounds that Mach and Planck shared (2010).

Like Planck, the second challenger acknowledged that he was addressing epistemological perspectives because of the increasing acceptance of Mach's views. He also addressed questions of time and space, but this time clearly because of the centrality he thought science should play in upholding materialism. Vladimir Lenin attacked Mach partly because of the emergence of Machian thought among socialist circles in both Zürich and Moscow (as well as Vienna). Having ourselves learned of Mach through his study of the conservation of work (which Lenin also quoted), we can see why many socialists would have integrated Mach's attacks on absolute time and space with the political and economic perspectives of socialism. Most important for Lenin was the view that Mach had slipped from relativistic thought to idealistic solipsism and that while Marx and Engels had developed a form of historical materialism, it was now necessary to reassert a materialist basis for epistemology in the objective reality of matter, independent of mind (Marot 1993). Mach never addressed Lenin's attack, but historians have shown that Mach remained an important but ambiguous resource for many in the Soviet Union, with the charge of being a Machian used as a weapon among different groups of scientists to enormous political effect in the 1920s and 1930s and also playing into their approach to Einstein (Bukharin 1931; Hessen 1931; Graham 1985; Staley 2011; Staley 2018).

Einstein becomes a relativist

Understanding this historical context and recognizing the enormous worldwide popularity that the theory of relativity won in 1919, it may seem surprising that the first time Einstein himself used the term "relativist" in print or correspondence seems to have been in addressing his opponents in 1918. Yet, this illustrates several important features of Einstein's physics and the way it became well known. The most remarkable aspect of Einstein's achievements in 1905 is the variety of the intellectual tools that he used, from techniques in statistical mechanics to a focus on a range of problematic and revealing phenomena like the photoelectric effect and Brownian motion, to an axiomatic approach based upon principles (and resolving tensions between them). With this plurality of approaches and an endeavor to approach the three different fields of mechanics, electrodynamics, and thermodynamics consistently, Einstein is unlikely to have identified himself closely with any single tool or principle. Further, he advanced the principle of relativity strongly aware of its initial limitation to inertial motions and in a period in which most German physicists (and Einstein himself) were writing interchangeably of the Lorentz theory of the electron, the Lorentz-Einstein theory, or the Lorentz-Einstein principle of relativity, even as they began to focus more strongly on Einstein's interpretation (Staley 2008, 301–309).

Spurred by his now-famous thought experiment on a man falling, in 1907 Einstein articulated the equivalence principle extending the principle of relativity from inertial motion to acceleration and gravitation (and he may have unconsciously drawn on Mach's earlier discussion of action and reaction and the significance of relative acceleration for a load at rest on a falling table) (Staley forthcoming 2019; Staley 2013). Revealingly, in 1914 Einstein wrote about the ensuing endeavor for a theory of gravitation as a tale of two relativity theories. The first, more narrow theory was founded on many experiments and was well accepted. The other was more speculatively based upon an epistemological demand Mach had first stated clearly, critiquing Newton (1914a, 337, 344). In Einstein's papers of this period, Mach is described as an "epistemologist," a term reflecting the title Mach had given his reply to Planck (1914b, V). On Mach's death in 1916, Einstein offered a moving obituary that began with a tribute to the significance of epistemology despite its poor reputation among subject specialists, and described how close Mach had been to demanding a general theory of relativity, almost fifty years earlier. He described Mach as a natural scientist and argued using the term "sensations" had led those unfamiliar with Mach's work to mistake the sober and careful thinker for a philosophical idealist and Sophist (1916, 101 (epistemology), 103 (general relativity), 104 (scientists, sensations)). When celebrating Planck's birthday in 1918, Einstein also touched on Planck's polemic, arguing that it stemmed from physicists' insistence that despite the in-principle possibility of alternatives, in practice a singular theoretical construction would prove itself superior at any time (and, remarkably, this could be related to a few basic laws in a Leibnizian "pre-established harmony" with the world of phenomena). This fueled the earnest reproach that some epistemologists valued this circumstance insufficiently (1918b, 31).

Thus, in the writings of Mach, Planck, Einstein, and others, the directions of a person's work might be summed up in the careful use of general terms like "physicist, " "epistemologist, " and "natural scientist. " Moving between early pluralities, Einstein wrote of adherents of the principle and the theory in diverse senses, but we should also observe the importance of singular perspectives for his view of theoretical physics and recognize that developing a consistent theory of gravitation in 1915 made it possible for Einstein to think in similarly singular terms of "general relativity." Generalized as this had been from "special relativity" and resting on the general covariance of the field equations in contrast to an earlier, unsuccessful effort, Einstein wrote that it would be barely possible for anyone who had really understood the theory to escape its charm, but although qualifying them variously, his technical papers always used locutions like "the theory of relativity" (1915, 779). While Mach had first used the more general and more personal term "relativist" when he felt that many others had joined his adherence to the significance of relative motion, it was opposition that called forth the personalized identity of the relativist for Einstein.

Writing for *Die Naturwissenschaften* in 1918, Einstein depicted a friendly debate between a persistent but polite Critic visiting a forthright Relativist. He conducted it fully on the grounds of physical theory, with a brief footnote drawing attention to the fact this was physical and not philosophical relativity. Focusing first on apparent paradoxes in the treatment of time, and later on an example that Philipp Lenard had raised about the different senses in which a jolt experienced by a railway car could be treated as an effect of the train or of the environment, Einstein's Relativist explained distinctions between equivalent and non-equivalent coordinate systems and special and general relativity. One key point is that the distinction between real and fictional gravitational fields is less important than that between quantities inherent in a physical system (independent of the choice of coordinate system) and those that do depend on the coordinate system; the Relativist also admits that relativity offers a less direct connection with measurement than previous theories (1918a).

Einstein staged this debate as a conversation among colleagues, without the pretensions of science writers or theatre critics. As well as the Nobel laureate Lenard, he addressed stances taken by Ernst Gehrcke, an optical expert at the *Physikalisch-Technische Reichsanstalt* who had attacked relativity while accusing Einstein of plagiarism (Wazeck 2009). Without naming Gehrcke, the Critic noted that he would not be accusing the Relativist of intellectual thievery; thus, Einstein raised questions about scientific decorum. His next article in *Die Naturwissenschaften* was a brief note in October 1919 reporting preliminary results of the eclipse expedition; his next account of relativity was written for the London *Times* a month later. Einstein's contribution to the media storm that surrounded his work featured a second and deeper, if humorous, nod to the significance of philosophical relativity.

Einstein's two-column article undoubtedly shaped public perceptions of him at the time and has also played a major role in scholarly understandings of his research. Opening with the significance of English science, scientists, and institutions in testing the results of a theory completed and published "in the country of their enemies," (1919b, 13) Einstein went on to articulate a distinction between constructive theories and theories of principle which has guided many interpretations of his work as well as that of the physics community more generally (see Seth 2008). He offered a brief account of the logical development of the two theories of relativity and underlined the role of Newton's work as a foundation "on which our modern conceptions of physics have been built," (Einstein 1919b, 14) a point that is usually regarded as a sop to British sensitivities but reflects longstanding views. Einstein finished with what he described as an "application of the theory of relativity" that would meet the taste of his readers. Prompted by the "amusing feat of the imagination" that (on 8 November) had led the *Times* to describe him as a "Swiss Jew" in contrast to the representation of him as a "German man of science" in Germany, Einstein anticipated the reversal of these descriptions should he ever be seen as a *bête noir* (Einstein 1919b, 14). The editors conceded his jest, commenting that Einstein had delivered no absolute description of himself ("Dr. Einstein's Theory," 1919).

Yet, Einstein's ironic distance should not lead us to discount the personal and intellectual basis for his quip linking nations, science, the status of Jews, and his own reputation. This had a deep and long-standing foundation in Einstein's attitude toward peoples and nations—and he sometimes pursued this by analogy to scientific progress, while also pointing to the limits of truth. Further, these stances rapidly became important to the way others perceived him, and very soon Einstein was being approached from the kind of divergent perspective he joked about. In October 1914, Einstein had joined the Berlin physiologist Georg Nicolai in signing a plea for European internationalism, even gesturing toward a new supra-state order (Nicolai et al. 1914; Rowe and Schulmann 2007). The manifesto gained only four signatures, but its sentiment was critical to Einstein. In a revealing manuscript written after he completed relativity, Einstein described science as necessarily built constructively upon the work of earlier generations, never destructively in the way a tyrannical ruler overthrows another. Relativity theory overthrew Newton's and Maxwell's theories just as little as a *Völkerbund* (league of nations) would demolish the states that chose to join it, for they would only have to modify their laws for the prize of higher security (Einstein, after December 1916, 5–6). His analogy, with its anticipatory hopes, remained unpublished, but as the war ended broader recognition of Einstein's stance against war was important to his reputation inside and outside of Germany, to diverse effect amidst chauvinism, and this was rapidly conjoined with his attitudes toward the situation of Jews in Germany, and Zionism.

Indeed, together with his *Times* article, two articles that Einstein published in Berlin newsprint offer a revealingly comprehensive perspective on how he hoped to work in the public as he first assumed the role of a public intellectual. In *Berliner Tagesblatt* on Christmas Day 1919, Einstein offered a short meditation on induction and deduction to show that devotion to eternal goals would serve political reconvalescence better than political deliberations or credos. Yet, one principal argument was that the complexity of scientific knowledge meant that while logical inconsistency or inability to meet the facts might be demonstrated, the *truth* of any given theory could never be proven. Einstein used theoretical under-determination to explain how sagacious scientists knowing fact and theory could still be passionate adherents of different theories (without the gloss on practical singularity he had given for Planck) (1919c). Einstein thus distinguished the goals of science, but not its passions or truth from politics, and himself stepped into the political fray five days later. In a forthright article, he argued that perceptions of the social and economic threat posed by the rise of "Eastern Jews" in Berlin among both Jewish and other Germans were economically unfounded, historically short-sighted in the light of previous migrations, and morally questionable (1919a). Einstein worked to show that sentiment among different groups of people reflected their social position and might be addressed by gaining a more comprehensive understanding of the different dimensions involved. Addressing the public in those early days, Einstein thus offered precise and subtle renderings of both theoretical physics and political concerns, and was ready to link them.

Within a short period of time, these issues had become still more personally entwined for Einstein in two related ways. The first concerned finding that an anti-relativity event in August 1920, in which Gehrcke participated, also sponsored anti-Semitism (van Dongen 2007). In Gehrcke, then, Einstein learned that he had already met a *bête noir* figure. Einstein's impassioned newspaper response opened another side of the thought he had joked of in Britain, but his ellipses show that he could not complete it in Germany—or that he wanted his readers to supply the thought themselves. Raising the possibility that arguments against relativity were shaped by motives other than a striving for truth, he wrote: "Were I a German patriot with or without a swastika instead of a Jew with liberal international views, then…" (Einstein 1920, 1). Einstein also promised that those who were ready to bring their objections before a scientific audience would have the chance to do so in a discussion at the annual meeting of German scientists and physicians in Bad Nauheim (Einstein 1920, 2). There, he met Lenard for the first time, and they went over much of the ground that Einstein had considered in his 1918 dialogue, with Lenard relating his objections to his emphasis on mechanical and intuitive images rather than mathematical treatments in physics, and Einstein responding that what counts as intuitive changes with time (and he thought physics was primarily conceptual) (Lenard et al. 1920). Reporting on the event, Hermann Weyl wrote that Lenard simply did not comprehend Einstein's theory—and notably embedded his one reference to "relativists" in quotation marks (1920a, 124; 1920b). Whether they were adherents or not, neither Einstein nor others had adopted the label of "relativist" widely.

In contrast, within a year Einstein was describing himself as a Zionist. This was a controversial stance in assimilationist Germany, and as he toured the United States to raise funds for Zionism, Einstein gave a lengthy interview that became the basis for two articles published under his name. Einstein said he had been unaware of his Jewishness until he moved to Berlin in 1914, when the plight of young Jews attempting to pursue their studies or to gain a foundation for existence in a strongly anti-Semitic environment became clear to him. Zionism for Einstein was a cultural necessity to increase the self-confidence of Jews—and this would also help them live normally with non-Jews—as much as it was a colonial movement in Palestine, and he argued that it did not preclude cosmopolitanism. Einstein outlined the factors that he thought shaped perceptions of anti-Semitism, Zionism, and his own attitudes in such detail that the German- and English-language versions bore revealingly different titles: " Wie ich Zionist wurde" ("How I Became a Zionist"), and "Jewish Nationalism and Anti-Semitism. Their Relativity," respectively.

The first version was explicitly reviewed by Einstein, while it is notable that *The Jewish Chronicle* (where the second appeared) readily referred to the relativity of knowledge. Similarly, Lord Haldane and James Jeans both extended such thoughts to nations and relativity theory in careful play when welcoming Einstein to Britain in June 1921 (Viscount Haldane 1921).

I noted earlier that historians of science have often emphasized the problem of confusing physical and philosophical relativity, citing the potential for other names that would draw attention to the invariants that the physical theory highlights. Like Einstein, however, we should go further than simply noting this difference and take his need to joke more seriously. I have shown here that Mach and Einstein were only briefly "relativists." Their different uses of this term reflect very different stages in the development of the technical tools and public profile of relativity, and came variously in response to the acceptance or denial of their work among professional peers. While recognizing the limitations of absolutes and mastering the conceptual and mathematical demands of special and general relativity were surely the central marks of being a relativist for Mach and Einstein, it is notable that each was also ready to work across the contested terrain of science and politics. Further, as Engels's work and Einstein's joke equally show, this terrain was already scored by another sense of the term that Mach and Einstein had chosen for their science, in the tensions between relative and absolute across scientific disciplines, ethics, and politics.

While Mach's attacks on absolutes proved attractive to some socialists, they also drew Lenin's misplaced ire as idealistic and solipsistic. Similar charges were later brought against Einstein's work by some Soviets, while conservatives in Germany sometimes treated his theory as Bolshevik. The problems posed by such opportunistic political attacks are surely one reason to emphasize the distinction between physical and philosophical relativism. Yet, as Mach argued for the historical contingency of scientific theory in order to change it, Einstein made philosophical relativity part of public discussion of his work, both in jest and in warning, in order to be able to point to the assumptions underlying responses to his work. Far from laughing at the taste of readers or simply denying the relevance of philosophical relativity, in choosing not to separate scientific from social relativity, both Mach and Einstein engaged their understanding of the social and political rather closely with their perspectives on science. Each sought to address both science and politics creatively—and they were right to recognize that they had to live and work with each of these forms of relativity.

Bibliography

Blackmore, J. T. (1972), *Ernst Mach: His Work, Life, and Influence*, Berkeley: University of California Press.

Bukharin, N. I. (1931), "Theory and Practice from the Standpoint of Dialectical Materialism," in *Science at the Cross Roads: Papers Presented to the International Congress of the History of Science and Technology held in London from June 29th to July 3rd, 1931 by the Delegates of the U.S.S.R.*, London: Kniga, 1–23.

"Dr. Einstein's Theory," (1919), *The London Times*, 28 November, 13.

Einstein, A. (1914a), "Zum Relativitätsproblem," *Scientia* 15: 337–348.

——. (1914b), "Zur Theorie der Gravitation," Naturforschende Gesellschaft, Zürich, *Vierteljahrsschrift* 59: 4–6.

—— (1915), "Zur allgemeinen Relativitätstheorie," Königlich Preußische Akademie der Wissenschaften, Berlin, *Sitzungsberichte*: 778–786.

—— (1916), "Ernst Mach," *Physikalische Zeitschrift* 17: 101–104.

—— (1918a), "Dialog über Einwände gegen die Relativitätstheorie," *Die Naturwissenschaften* 6: 697–702.

—— (1918b), "Motive des Forschens," in *Zu Max Plancks sechzigstem Geburtstag. Ansprachen, gehalten am 26. April 1918 in der Deutschen Physikalischen Gesellschaft von E. Warburg, M. v. Laue, A. Sommerfeld und A. Einstein*, Karlsruhe: C.F. Müllersche Hofbuchhandlung, 29–32.

—— (1919a), "Die Zuwanderung aus dem Osten," *Berliner Tageblatt* (Morgen-Ausgabe), 30 December, 2.

—— (1919b), "Einstein on his Theory. Time, Space and Gravitation. The Newtonian System," *The London Times*, 28 November, 13–14.

—— (1919c), "Induktion und Deduktion in der Physik," *Berliner Tageblatt* (Morgen-Ausgabe), 25 December, 1.

—— (1920), "Meine Antwort. Ueber die anti-relativitätstheoretische G. m. b. H.," *Berliner Tageblatt* (Morgen-Ausgabe), 27 August, 1–2.

—— (1993), *The Collected Papers of Albert Einstein*, Vol. 5, *The Swiss Years: Correspondence, 1902–1914*, edited by Martin J. Klein, A. J. Kox and Robert Schulmann, Princeton: Princeton University Press.

—— (after December 1916), "Die hauptsächlichen Gedanken der Relativitätstheorie," in Janssen, M., R. Schulmann, J. Illy, C. Lehner & D. K. Buchwald (eds.) *The Collected Papers of Albert Einstein*, Vol. 7: *The Berlin Years: Writings, 1918–1921*, Princeton: Princeton University Press, 2002, 3–7.

Engels, F. (1878), *Herrn Eugen Dühring's Umwälzung der Wissenschaft: Philosophie, Politische Oekonomie, Sozialismus*, Leipzig: Verlag der Genossenschafts-Buchdruckerei.

Fechner, G. T. (1860), *Elemente der Psychophysik*, Vols. 1 and 2, Leipzig: Breitkopf und Härtel.

Graham, L. R. (1985), "The Socio-Political Roots of Boris Hessen: Soviet Marxism and the History of Science," *Social Studies of Science* 15: 705–722.

Heidelberger, M. (2004), *Nature from Within: Gustav Theodor Fechner and his Psychophysical Worldview*, Pittsburgh: University of Pittsburgh Press.

Hentschel, K. (1990), *Interpretationen und Fehlinterpretationen der speziellen und der allgemeinen Relativitätstheorie durch Zeitgenossen Albert Einsteins*, Science Networks Historical Studies, Basel/Boston: Birkhäuser.

Herbert, C. (2001), *Victorian Relativity: Radical Thought and Scientific Discovery*, Chicago: University of Chicago Press.

Hessen, B. (1931), "The Social and Economic Roots of Newton's 'Principia'," in *Science at the Cross Roads: Papers Presented to the International Congress of the History of Science and Technology Held in London from June 29th to July 3rd, 1931 by the Delegates of the U.S.S.R.*, London: Kniga, 1–62.

Lenard, P., A. Einstein, C. Rudolph, M. Palágyi, M. Born, G. Mie and O. Kraus (1920), "Allgemeine Diskussion über Relativitätstheorie," *Physikalische Zeitschrift* 21: 666–668.

Mach, E. (1860), "Über die Änderung des Tones und der Farbe durch Bewegung," *Sitzungsberichte der Mathematisch-naturwissenschaftlichen Classe der Kaiserlichen Akademie der Wissenschaften Wien* 41: 5435–5460.

—— (1861), "Über die Controverse zwischen Doppler und Petzval, Bezüglich der Änderung des Tones und der Farbe durch Bewegung," *Zeitschrift für Mathematik und Physik* 6: 1201–1226.

—— (1862), "Über die Änderung des Tones und der Farbe durch Bewegung," *Annalen der Physik und Chemie* 116: 3333–3338.

—— (1863), "Vorträge über Psychophysik," *Oesterreichische Zeitschrift für praktische Heilkunde* 9: 146–148, 167–170, 202–204, 225–228, 242–245, 260–261, 277–279, 294–298, 316–318, 335–338, 352–354, 362–366.

—— (1868), "Ueber die Definition der Masse," *Repertorium für physikalische Technik, für mathematische und astronomische Instrumentenkunde* (Carl's Repertorium der Physik) 4: 355–359.

—— (1872), *History and Root of the Principle of the Conservation of Energy*, translated by P. E. B. Jourdain, Chicago/London: Open Court/Kegan Paul, Trench, Trübner & Co, 1911.

—— (1883), *Die Mechanik in ihrer Entwickelung historisch-kritisch dargestellt*, Leipzig: F.A. Brockhaus.

—— (1886), *Beiträge zur Analyse der Empfindungen*, Jena: Fischer.

—— (1889), *Die Mechanik in ihrer Entwickelung historisch-kritisch dargestellt*, 2. verbesserte Auflage, Leipzig: F.A. Brockhaus.

—— (1897), *Die Mechanik in ihrer Entwickelung historisch-kritisch dargestellt*, 3. verbesserte u. vermehrte Auflage, Leipzig: F.A. Brockhaus.

—— (1901), *Die Mechanik in ihrer Entwickelung historisch-kritisch dargestellt*, 4. verbesserte u. vermehrte Auflage, Leipzig: F. A. Brockhaus.

—— (1908), *Die Mechanik in ihrer Entwickelung historisch-kritisch dargestellt*, 6. verbesserte u. vermehrte Auflage, Leipzig: F. A. Brockhaus.

—— (1910), "Die Leitgedanken meiner naturwisseschaflichen Erkenntnislehre und ihre Aufnahme durch die Zeitgenossen," *Physikalische Zeitschrift 11*: 599–606.

—— (1942), *The Science of Mechanics, a Critical and Historical Account of Its Development*, translated by T. J. McCormack, 5th ed. La Salle, IL/London: Open Court.

Marot, J. E. (1993), "Marxism, Science, Materialism: Toward a Deeper Appreciation of the 1908–1909 Philosophical Debate in Russian Social Democracy," *Studies in East European Thought* 45: 147–167.

Miller, A. I. (1981), *Albert Einstein's Special Theory of Relativity: Emergence (1905) and Early Interpretation (1905–1911)*, Reading, MA: Addison Wesley.

Nicolai, G. F., A. Einstein and W. J. Förster (1914), "Aufruf an die Europäer," *The Collected Papers of Albert Einstein*, Vol. 6: *The Berlin Years: Writings, 1914–1917*, edited by A. J. Kox, Martin J. Klein and Robert Schulmann, Princeton: Princeton University Press, 1996, 69–71.

Planck, M. (1909), "Die Einheit des physikalischen Weltbildes," *Physikalische Zeitschrift* 10: 62–75.

Rowe, D. E. and R. Schulmann eds. (2007), *Einstein on Politics: His Private Thoughts and Public Stands on Nationalism, Zionism, War, Peace, and the Bomb*, Princeton: Princeton University Press.

Schuster, P. (2005), *Moving the Stars: Christian Doppler, His Life, His Works and Principle, and the World After*, Pöllauberg, Hainault: Living Edition.

Seth, S. (2008), "Crafting the Quantum: Arnold Sommerfeld and the Older Quantum Theory," *Studies in History and Philosophy of Science* 39: 335–348.

Staley, R. (2008), *Einstein's Generation: The Origins of the Relativity Revolution*, Chicago: University of Chicago Press.

—— (2011), "Culture and Mechanics in Germany, 1869–1918: A Sketch," in Carson, C., A. Kojevnikov & H. Trischler (eds.) *Weimar Culture and Quantum Mechanics: Selected Papers by Paul Forman and Contemporary Perspectives on the Forman Thesis*, London/Singapore: Imperial College Press/World Scientific Publishing, 277–292.

—— (2013), "Ernst Mach on Bodies and Buckets," *Physics Today* 66 (12): 424–427.

—— (2018), "The Interwar Period as a Machine Age: Mechanics, The Machine, Mechanisms and the Market in Discourse," *Science in Context* 31:3 (2018) 263–292.

—— (Forthcoming 2019), "Revisiting Einstein's Happiest Thought: From the Physiology of Perception to Experimental Propositions and Principles in the History of Relativity," in F. Stadler (ed.), *Ernst Mach—Life, Work, and Influence / Ernst Mach—Leben, Werk und Wirkung Vienna Circle Institute Yearbook*, Dordrecht: Springer.

Stallo, J. B. (1881), *The Concepts and Theories of Modern Physics*, edited by Percy W. Bridgman, Cambridge, MA: Belknap Press 1960.

van Dongen, J. (2007), "Reactionaries and Einstein's Fame: "German Scientists for the Preservation of Pure Science," Relativity, and the Bad Nauheim Meeting," *Physics in Perspective* 9 (2): 212–230.

Viscount Haldane, R. (1921) "Professor Einstein, Visit to England. Brilliant Reception. Lord Haldane's Tribute," *The Jewish Chronicle*, 17 June, 26.

Wazeck, M. (2009), *Einsteins Gegner. Die öffentliche Kontroverse um die Relativitätstheorie in den 1920er Jahren*, Frankfurt am Main: Campus.

Wegener, D. (2010), "De-Anthropomorphizing Energy and Energy Conservation: The Case of Max Planck and Ernst Mach," *Studies in History and Philosophy of Modern Physics* 41: 146–159.

Weyl, H. (1920a), "Antwort auf Prof. Dr. Gehrcke," *Die Umschau* 25: 1231–1234.

—— (1920b), "Die Diskussion über die Relativitätstheorie," *Die Umschau* 24: 609–611.

Wolters, G. (2012), "Mach and Einstein, or, Clearing Troubled Waters in the History of Science," in Lehner, C., J. Renn & M. Schemmel (eds.), *Einstein and the Changing Worldviews of Physics*, Berlin: Springer, 39–57.

Ziche, P. G. (2008), "Fechner und die Folgen außerhalb der Naturwissenschaften," *Scientia poetica* 12: 376–389.

6   Husserl on relativism

Dermot Moran

Introduction

The Moravian-born, German-speaking founder of phenomenology, Edmund Husserl (1859–1938) had a life-long engagement with relativism, from his *Logical Investigations* (1900–1901) to his late *Crisis* writings (1934–1937) (see Soffer 1991; Carr 1987; Mohanty 1997). Furthermore, he maintained a singularly consistent critique of relativism from the beginnings of his career, when he targeted *psychologism* as a relativistic threat to the objectivity of logic and of scientific knowledge more generally, right through to the end of his career. Thus, in late essays such as "Phenomenology and Anthropology" (1931, Husserl 1927–1931), he portrayed modern subjectivism as moving in the directions of a relativistic "anthropologism" (Dilthey, Heidegger), as opposed to a true science of subjectivity, which he, of course, claimed to have initiated with his transcendental phenomenology. In this sense, then, Husserl was the first person to explicitly expose the relativism inherent in Heidegger's attempt to found phenomenology on *Dasein*. (Later writers such as Richard Rorty have interpreted Heidegger as a relativist; see also Lafont 2000.)

Husserl tends to identify relativism with skepticism, so he frequently talks about "skeptical relativism" (as in his *First Philosophy* lectures 1923/24, 66). The mature Husserl wanted an entirely new science of transcendental subjectivity, which, according to his claims, would ground all the other sciences, not just the natural sciences, but also the human sciences, including psychology, history, and all other so-called *Geisteswissenschaften* in an absolute way that would banish forever the threat of relativism and skepticism. Thus, he writes in the *Crisis of European Sciences*, philosophy has not yet become truly scientific, i.e., transcendental, and the modern discovery of subjectivity collapsed into psychological subjectivity and hence into "anthropologistic relativism" (1934–1937, 69). True philosophy, operating with a transcendental vigilance provided by the *epoché*, is necessary to prevent relapse into objectivism, naturalism, anthropologism, and naïveté and to offer a true meaning-clarification of the nature of the world as a *Leistung* (achievement) of transcendental subjectivity. Husserl believed that transcendental phenomenology made it possible to offer, for the first time, a true science of the life-world in which people always live and which the positive sciences take for granted. While there is an undoubted plurality of life-worlds, there is, from the transcendental point of view, also the possibility of a universal science of the life-world (1934–1937, 123ff.) that uncovers in a new way the global correlation between modes of givenness and modes of subjective apprehension.

In his early years, Husserl, as a mathematician and logician, was a strong defender of the fixed and unchangeable nature of ideal truths and hence an opponent of all forms of relativism and subjectivism. Initially, his main target was the psychologistic approach to logic, which he found in many nineteenth-century logicians (not just in empiricists such as J. S. Mill, but also in German logicians and psychologists such as Christoph von Sigwart, Benno Erdmann, Wilhelm Wundt, and others, see Kusch 1995), but later, especially in his essay "Philosophy as Rigorous Science" (1910–1911) he expanded this critique to all forms of reductive naturalism. He was also concerned that the human sciences, especially the historical sciences, were also accepting a form of relativism in so far as they had a tendency toward *historicism*, the view that historical periods had specific *Weltanschauungen* (world-views) or outlooks that could only be understood or appreciated from within, thus making an objective science of history as such impossible, since validity and truth would be relative to a particular era and its outlook. In this regard, Husserl specifically targeted the work of Wilhelm Dilthey (who himself, however, in a letter to Husserl in June 1911, shortly before he died, rejected the charge of scepticism and relativism; Dilthey 1911; Bambach 1995, 173). In his later years, in part due to his contact with Martin Heidegger in Freiburg, Husserl took the historicity of human existence much more seriously and recognized that peoples belong to different life-worlds and that their cultures have their own *Geschichtlichkeiten* (specific historicities). Nevertheless, while acknowledging the plurality of life-worlds and the fact that a certain cultural *relativity* is inevitably present, he sought a universal science that exposed the fundamental features shared by all life-worlds. In this sense, Husserl distinguished—although not in any explicitly thematic way—between relativity as an empirical fact or truism of culture and relativism (which is a philosophical theory about the nature of truth).

The critique of psychologism as leading to relativism (1900)

At the beginning of the twentieth century, Husserl correctly predicted that naturalism and relativism would be two of the greatest philosophical movements of the twentieth century. He regarded both as threats to the true nature of philosophy as rigorous science. In fact, Husserl's insight that all psychologistic interpretations of logic would inevitably lead to relativism (and thence inevitably to skepticism) was already the driving force behind the first volume of his *Logical Investigations*, the *Prolegomena to Pure Logic*, published in 1900.[1] In this *Prolegomena*, Husserl also recognized several kinds of relativism, including what he called "species relativism" or "anthropologism," according to which what is logically valid is relative to what the human mind is capable of understanding or what laws govern the human mind in its actual operation. For Husserl, this species relativism was particularly to be found among neo-Kantians. He also diagnosed the tendency to reduce the natural sciences to the laws of human psychology as found in various psychological thinkers (e.g., Mill) as equally dangerous.

In order to overcome relativism and skepticism, the *Prolegomena* mounted a strong defense of the conception of truth as ideal and universal and "true for all." Husserl claimed to be inspired by Bolzano and Lotze to recognize that "truths in themselves" or "propositions in themselves" are essentially objective states of affairs. In this regard, Husserl specifically rejected the relativist claim (later endorsed by Heidegger) that the truth of scientific laws is relative to the era in which they are discovered, i.e., that Newton's laws, strictly speaking, were not true before Newton formulated them. Heidegger states in *Being and Time* § 44:

> Newton's laws, the principle of contradiction, any truth whatever—these are true only as long as *Dasein* is. Before there was any *Dasein*, there was no truth; nor will there be any after *Dasein* is no more.
>
> (1927, 269)

Husserl argues, on the contrary, that the reality, e.g., of gravity, specified in Newton's Laws, holds irrespective of whether these laws were formulated or not, or whether they ever were contemplated by human minds, although laws of course may be formulated in different ways with different degrees of precision.

The *Prolegomena to Pure Logic* aimed to secure the true meaning of logic as a pure, *a priori*, science of ideal, objective meanings and of the necessary formal laws regulating them, entirely distinct from all contingent psychological acts, contents, and procedures. The *Prolegomena* discussed at length various *psychologistic* interpretations of logic, propounded by John Stuart Mill and others (Husserl's list includes Bain, Wundt, Sigwart, Erdmann, and Lipps; 1900/1901 **I**, 83), which Husserl viewed as leading to a skeptical relativism that threatened the very possibility of objective knowledge. Turning instead to an older tradition of logic that he traces to Leibniz, Kant, Bolzano, and Lotze, Husserl defends a vision of logic as a pure *Wissenschaftslehre* (theory of science)—in fact, the "science of science,"—in the course of which he carefully elaborates the different senses in which this pure logic can be transformed into a normative science or developed into a practical discipline or technology (*Kunstlehre*).

Husserl himself regarded his *Logical Investigations* as a "breakthrough work (*ein Werk des Durchbruchs*) not an end but rather a beginning" (1900/1901 **I**, 3). It certainly contained many of the arguments he would rehearse again through his career. As Husserl put it, the *Investigations* originally grew out of his desire to achieve "a philosophical clarification of pure mathematics" (1900/1901 **I**, 1). In the *Prolegomena*, Husserl explicitly abandoned his own earlier approach to logic and mathematics expressed in his first book, *Philosophy of Arithmetic* (1891), which had been judged psychologistic by its chief critic Gottlob Frege (1848–1925). In his review of Husserl's *Philosophy of Arithmetic* Frege had sharply criticized it, and Husserl came to agree with Frege's criticism, although he had already diagnosed himself the problems with his own earlier approach. In his correspondence with Frege, he agrees that both are in search of ideality (see Mohanty 1982; Hill and Haddock 2000). In his *Philosophy of Arithmetic*, Husserl had given an account of the genesis of arithmetic concepts using Brentanian descriptive psychology. Indeed, he claimed to have been particularly drawn to Brentano's project for a reform of Aristotelian logic in Brentano's 1884–1885 lecture-course, *Die elementare Logik und die in ihr nötigen Reformen* (Elementary Logic and its Necessary Reform) which Husserl attended in Vienna. As Husserl recalled in his 1919 memorial essay for Brentano: "Brentano's pre-eminent and admirable strength was in logical theory" (1919, 345). Husserl seemed especially interested in, and critical of, Brentano's novel structure of judgments and his construal of judgment as assertion or denial of an object. Indeed, Husserl—no doubt exaggerating somewhat—later presented his own *Logical Investigations* as an attempt to do justice to the extraordinary genius of Brentano by overcoming the latter's psychologistic grounding of logic. In the *Philosophy of Arithmetic*, Husserl proposed "psychological analyses" in the Brentanian sense. Husserl relies heavily on Brentano's distinction between physical and psychical relations to argue that the way we group items together in order to count them requires grasping higher-level "psychical" or "meta-physical" relations between the items, as opposed to the more usual "primary" or "content" relations. By the time of the *Prolegomena*, Husserl was intent to distance himself from any psychological grounding of arithmetic or logic and to defend the objectivity and ideality of logical and mathematical truths and laws and the objects they governed over. Clearly, now Husserl thought that a purely *psychological* grounding of logic in human thought processes inevitably ended in relativism.

Instead, Husserl retrieved "pure logic," a conception found in Leibniz and Kant, but expressed most clearly in *Theory of Science* (Bolzano 1837) of the neglected Austrian logician Bernard Bolzano (1781–1848), and his followers (especially Frege's teacher Rudolf Hermann Lotze, 1817–1881). This tradition treated logic as a purely formal "science of science," and recognized that judgments, statements, expressions, or "propositions" articulated in language, if true, mirrored ideal states of affairs. Thus, the state of affairs expressed in the so-called Pythagorean Theorem—the square of the hypotenuse of a right triangle is equal to the sum of the squares of the other two sides—stands as an independently valid truth, whether or not anyone actually thinks it or expresses it in language, or whether it is in fact ever discovered by any cognizing subject. Such propositional states of affairs—perhaps mislead-ingly called "thought contents"—possess an "ideality" that allows them to be instantiated in different thought processes of the same individual (2001, **I**, 167) or in diverse individuals' thoughts at different times.

Chapter Seven of the *Prolegomena*, entitled "Psychologism as a Sceptical Relativism," offers a comprehensive refutation of psychologism. Husserl begins by saying that the worst objection that can be made against any theory is that it does not conform to the requirements of what a theory as such should be:

> The worst objection that can be made to a theory, and particularly to a theory of logic, is that it goes against the self-evident conditions for the possibility of a theory in general.
>
> (2001, **I**, 75)

Husserl goes on to distinguish two kinds of conditions of possibility—objective and subjective. Subjective conditions are not here the limitations of any particular psychological subject but the conditions determining what counts as knowledge as such, whereas objective conditions pertain to the manner in which propositions relate to one another in terms of ground and consequence.

Prolegomena section 34 lays out "The Concept of Relativism and Its Specific Forms." Husserl distinguishes two kinds of relativism: what he calls "individual" relativism, and "species" or "specific" relativism, one particular version of which he calls "anthropologism." Anthropologism is the claim that "all truth has its source in our common human constitution" (2001, **I**, 80). Husserl distinguishes between skepticism and relativism. He sees skepticism as a more general metaphysical theory about the possibility of knowledge as such, and, since in the *Logical Investigations* he is not interested in such metaphysical theories (§ 33), he is more interested in skeptical challenges to the possibility of scientific knowledge, and this leads him to see skepticism as inevitably leading to relativism.

Husserl initially introduces "subjectivism" or "relativism" in terms of Protagoras' claim that "man is the measure of all things" (2001 **I** § 34). Husserl then interprets relativism in more or less the manner of Protagoras:

> "The individual man is the measure of all truth." For each man that is true what seems to *him* true, one thing to one man, and the opposite to another, if that is how he sees it. We can therefore opt for the formula "All truth (and knowledge) is relative—relative to the contingently judging subject."
>
> (2001 **I**, 77)

In this regard, all truth is relative to the contingent cognizing subject.

However, there is another kind of relativism that makes truth relative—not to the contingent subject but to the particular animal species—and in the case of humans this kind of species-relativism Husserl calls "anthropologism."

Husserl sees Kant's followers (Sigwart, Erdmann) as guilty of anthropologism in that for them the limits of knowledge are the limits of the human condition. Husserl is particularly critical of Sigwart, whom he claims argued that for a proposition to be true it must have been thought by someone.

Husserl takes the general view that relativism is self-refuting and amounting to *Widersinn* (countersense). To assert that relativism is true is to be committed to one absolute truth; hence, to *assert* relativism is *eo ipso* to refute it. Husserl acknowledges that the supporter of relativism will not be dissuaded by this argument because the relativist asserts that he is merely expressing his own standpoint (2001 **I**, 78). Husserl, however, thinks there is an inner contradiction in relativism in that it presupposes an objective standard of truth. All theories that make claims about reality assume the standpoint of an objective reality. Husserl accuses the neo-Kantians of "species relativism" or "anthropologism."

Anthropologism maintains is not relative to the individual but to the human species. As Husserl puts it, such a dependence can only be thought of as causal, and thus the human constitution would be *causa sui* in respect of its laws of truth. But if the human species were destroyed, and there were no longer a human constitution—would that mean that truth as such would disappear? Husserl thinks this is absurd (2001 **I**, 81). Husserl, on the contrary, is a "logical absolutist." For him, "every judgment is bound by the pure laws of logic without regard to time and circumstances, or to individuals and species" (2001 **I**, 93).

In the *Prolegomena* Husserl takes a strongly realist view of scientific truths ("realist" in the sense that truths are true independently of their being known). Newton's Law of Gravity, if true, was true even before Newton discovered it. To claim otherwise (as Sigwart did) would be self-contradictory for Husserl (2001 **I**, 85). That said, empirical scientific formulations, such as Newton's Law, for Husserl, do not have the universality and ideality of *a priori* laws. They are empirical approximations, capable of refinement, or indeed reformulation in more exact ways. Psychological laws, similarly, like the empirical laws of the natural sciences, are not the same as the *überempirischen und absolut exakten Gesetze* (meta-empirical and absolutely exact laws) of logic (§21, 2001 **I**, 48). Husserl also distinguishes between conceivability (or imaginability by humans) and truth itself. According to Husserl's realism, certain truths may never be discoverable by the human mind, but they are true nonetheless. Furthermore, psychological impossibility does not mean logical impossibility. Husserl thinks it may very well be possible for humans to deny the Principle of Non-Contradiction, but that does not mean the principle does not hold or is not independently valid.

Naturalism as a kind of relativism

In later writings, right down to "The Origin of Geometry" (1939), Husserl continued to defend the ideality and indeed univocity of genuine logical and mathematical truths. There is only one Pythagorean Theorem, and its validity is independent of the cultural or scientific norms prevalent at any time in history. Nonetheless, Husserl became more concerned that psychologism was not just a tendency deeply embedded in the psychology, logic, and mathematics of his day, but that it in fact formed part of a larger outlook—naturalism—that suffered from the same inherent self-contradictory and countersensical character. In his "Philosophy as Rigorous Science" essay, Husserl spoke of the "battleground of psychological naturalism" (1910/1911, 278) and later, in 1915, he wrote to Rickert that he was in agreement with the neo-Kantians in the struggle against "the naturalism of our time as our common enemy" (1915, 178).

Especially after 1906–1907, the years in which he discovered the *epoché* and the "reduction," Husserl expanded his target from psychologism to naturalism. His transcendental turn was in part a rejection of scientific objectivism and naturalism. For Husserl, naturalism involves a countersense. It assumes the validity of logic and mathematics, which it cannot find in nature. Naturalism is linked to what Husserl calls "the discovery of nature" (1910/1911, 252). Nature is posited by natural science as a domain subject to universal laws. However, Husserl maintained in his "Philosophy as Rigorous Science," "all natural science is naive by virtue of its starting point. The nature into which it wants to inquire is simply there for it" (1910/1911, 257). Husserl believed, furthermore, that a purely natural scientific epistemology was a countersense (1910/1911, 259). But for Husserl, "naturalism" is an inevitable consequence of an absolutization of the belief in the world inherent in what he called "the natural attitude." Naturalism naively takes as real what in fact is the way things are given under the natural attitude (Moran 2008).

Another form of relativity is introduced by Husserl once he discovers the notion of *natürliche Einstellung* (natural attitude, a notion that is first discussed in print in Husserl 1913 § 27, but that was already present as an idea in his 1906–1907 lectures) and its correlated "transcendental attitude." Husserl now argues that ontology is relative to the stance one adopts toward the world. Attitudes can be relative. Only the transcendental attitude can claim absoluteness. This becomes developed in Husserl's mature works, especially in his *Amsterdam Lectures* (of 1928; Husserl 1927–1931) and in the *Crisis of European Sciences*.

Historicism as relativism in "Philosophy as Rigorous Science" (1910/1911)

In his journal article "Philosophy as a Rigorous Science" (1910–1911), Husserl was concerned that relativism had crept not just into pure *a priori* sciences such as logic and mathematics but also into the *human* sciences. In this case, the particular tendency that he regarded as dangerous Husserl labelled "historicism." Husserl traces this tendency to post-Hegelian philosophy, which claimed that philosophies were true for their own time and had abandoned the overarching framework of absolute knowledge that Hegel had maintained. Husserl writes:

> Hegelian philosophy had after-effects due to its doctrine of the relative legitimacy of each philosophy for its time—a doctrine whose sense, of course, differed completely in a system that pretended to absolute validity from the historicistic sense in which the doctrine was adopted by the generations that had lost the belief not only in Hegelian philosophy but in any absolute philosophy whatsoever.
>
> (1910/1911, 252)

This gave rise to what Husserl called *Weltanschauungsphilosophie* (worldview-philosophy).[2] According to Husserl, Dilthey, with his account of the different kinds of worldview, was dangerously close to the view that each era could only be understood from within its prevailing worldview. This seemed to rule out the possibility of historical understanding *across* eras or epochs. If historicism is true, then the historians of the twentieth century could never fully understand the mind-set of the ancient Greeks. For Husserl, to concede such a claim would be the death of history as a strict science.

Historicism, for Husserl, does not explicitly *naturalize* spirit in the manner that contemporary scientific psychology does, but it does fall prey to relativism (1910/1911, 278) and for reasons closely analogous to the ones Husserl deploys against psychologism. Husserl writes: "The worldview-philosophy of modernity is ... a child of historicistic skepticism" (1910/1911, 283). Husserl agrees with Dilthey that it is a great task to study the morphology and typology of spiritual forms. In this regard, he refers to Dilthey's study of the *Typology of Worldviews* (Dilthey 1919; Husserl 1910/1911, 280). However, Husserl explicitly invokes the "relativity of the historical form of life," quoting Dilthey, as one form of life succeeds another, there is no claim to the absolute validity of one form of life. Of course, it is a factual truth that cultural forms are bound to their era and its prevalent outlook. But the tendency of this approach is to deny to any particular era an absolute validity. Husserl asks whether the assumption of a lack of absolute validity follows from the plurality of cultural forms:

> Certainly, a worldview and a worldview philosophy are cultural formations that come into being and disappear in the stream of the development of mankind, whereby their spiritual content is determinately motivated under the given historical circumstances. Yet the same holds also of the rigorous sciences. Do they for that reason lack objective validity?
>
> (Husserl 1910/1911, 280)

In a footnote, Husserl acknowledges that Dilthey himself did actually reject historicist skepticism. Nevertheless, he fails to see how Dilthey can consistently maintain his position given his position on worldviews.

Social unities of all kinds bear similarities to the world of organisms: everything is in development; there are no fixed species. We understand the life of spirit only by immersing ourselves in its motivations. Husserl accepts the factual truth of what Dilthey is asserting, but he questions its "legitimacy." Worldviews come and go—so also do sciences (mathematical theories may rise and fall)—but this does not undermine its objective legitimacy. According to worldview philosophy, the vast array of formations of historical consciousness rules out the claim of any one of them having absolute validity. Husserl acknowledges the possibility of an extreme relativism that would hold that different scientific theories will be valid at different times:

> A very extreme historicist might well affirm this, pointing here to the change in scientific views, how what is regarded today as proven theory will tomorrow be seen to be void, how some speak of certain laws, whereas others call them mere hypotheses, and still others call them vague notions.
>
> (1910/1911, 280)

Historicism carried through consistently, however, will end up in "extreme skeptical subjectivism":

> The ideas "truth," "theory" and "science" would then, like all ideas, lose their absolute validity.
>
> (1910/1911, 280)

Husserl is concerned that historicism would undermine the very idea of objective validity, a concept required by the ideal of science itself. On the historicist view, for an idea to have validity would simply mean it was a factual production of a particular time in the life of spirit. Husserl's solution is to distinguish between science *as cultural achievement* and science *as the system of valid theory*. Decisions about validity and normativity are not matters for the empirical sciences to decide. A distinction must be made between what obtains factually and what is valid. No human science can argue for or against validity claims that are never factual claims at all. The inference to historical relativism and skepticism (that no historical era has produced an era-transcending truth) is not just invalid, it is, for Husserl, a countersense, like 2 x 2 = 5 (1910/1911, 282). Furthermore, the historical untenability of a particular claim has nothing to do with its validity or invalidity. Just as a mathematician would never draw an inference about validity in mathematics from the history of mathematics, neither should the cultural scientist or historian. Husserl, however, concedes that he is not dismissing history:

> If I therefore regard historicism as an epistemological aberration that, owing to its countersensical consequences, must be just as brusquely rejected as naturalism, then I would nevertheless like to emphasize

expressly that I fully acknowledge the tremendous value of history in the broadest sense for the philosopher.

(1910/1911, 283)

Husserl defends not just the value of history but the possibility that discoveries in wisdom can be gleaned from the critical interrogation of worldviews (Moran 2011). But wisdom is an ideal, a value, a goal lying in the infinite, which has to be apprehended in a different way from any factual experience. Husserl concludes:

> The "idea" of worldview is accordingly for each age a different one, as should be quite clear from the foregoing analysis of its concept. By contrast, the "idea" of science is a supratemporal one, and here that means that it is not limited by any relation to the spirit of an age.
>
> (1910/1911, 287)

The elderly Dilthey (who died on 1 October 1911) was put out that he was the focus of Husserl's attacks and wrote to Husserl denying the charge of relativism (Dilthey 1994, 44–47). Years later, in his 1925 lectures, Husserl made amends, acknowledging Dilthey's important contribution to descriptive psychology (Husserl 1910/1911; 1925).

The relativity of life-worlds and the absolute transcendental science

Once Husserl introduces the notion of the *Lebenswelt* (life-world), around 1919, he recognizes the plurality of worlds and the fact that truths and norms are often relative to worlds. Husserl's *Crisis* continues to be relevant because it challenges philosophers and scientists to think about the nature of the present age with its dominant scientific and technological world view, that has led, as Husserl believed, to universalization but at the same time to a kind of flattening out of reason that has left many core human values unsupported and threatened. In his analyses of the current state and hegemony of the scientific-technological attitude, Husserl predicted the rise of naturalism, relativism, and irrationalism in the face of the dominant instrumental reason. Somewhat in the spirit of Nietzsche, Spengler and others, Husserl also was attentive to the general mood of *weariness* sweeping through Western culture in this crucial period of the 1930s. As he saw it, "Europe's greatest danger is weariness" (1934–1937, 299). There is a danger of "despair," of loss of sense of values, leading to estrangement, cultural collapse and, ultimately, to "barbarism." This collapse is occasioned by skeptical relativism. Husserl defends the redeeming universality of philosophy: "from the ashes of great weariness, will rise up the phoenix of a new life-inwardness (*Lebensinnerlichkeit*) and spiritualization as the pledge of a great and distant future for man, for the spirit alone is immortal" (1934–1937, 299). For Husserl, transcendental phenomenology is the science that grasps the intrinsic meaning and inner rationality of the *accomplishment of spiritual life* in all its forms. As Husserl proclaims in the "Vienna Lecture":

> The spirit, and indeed only the spirit, exists in itself and for itself, is self-sufficient (*eigenständig*); and in its self-sufficiency, and only in this way, it can be treated truly rationally, truly and from the ground up scientifically.
> (1934–1937, 297)

These optimistic reflections by no means disguise Husserl's acute awareness of the difficulties and complexities facing the contemporary philosophy who seeks to be a "functionary" in the service of humankind.

Husserl is challenging the cultural dominance of *scientism* (with its commitment to what he calls "objectivism") and *naturalism*, which he sees as having also led to the acceptance of varieties of cultural relativism and ultimately to skepticism. As a result, European intellectual culture in its highest achievement (i.e., the sciences) is threatened by a profound and growing *irrationalism*. Husserl's proposed solution involves first and foremost *Klärung* (clarification) of what exactly has happened through a transcendental reflection on the meaning of the modern scientific achievement (and its implications for the development of modern philosophy).

In the *Crisis*, Husserl argues forcefully that the undoubted fact of the relativity of living in a life-world, which changes with different cultures and historical trajectories ("historicities"), does not lead to relativism. In fact, Husserl always praises the Greek skeptics for recognizing the relativity of all experience, but their mistake was to conclude to relativism about truth. Husserl acknowledges that pre-scientific experience has its own relativities. He writes, "What is actually first is the 'merely subjective-relative' intuition of prescientific world-life" (1934–1937, 125). But this "subjective-relative" experience must not be dismissed in the name of a naïve scientific objectivity (as indeed happened in the development of modern science since Galileo). Rather, "the life-world is a realm of original self-evidences" (1934–1937, 127).

Husserls's reading of Lévy-Bruhl's account of "primitive" worlds

Despite Husserl's negative attitude toward what he called "anthropologism" and his belief that all forms of empirical psychology and social anthropology were naïve regarding their acceptance of the world, he himself was growing increasingly interested in issues of human culture and history, and what he called "generativity" (Steinbock 1995), i.e., the process of cultural development and change across history, especially with regard to inter-generational transmission. He was also attempting to understand the relation between his transcendental phenomenology and historical studies of human culture, and even, as this letter attests, delving into ethnological literature. Late in his own career, Husserl encountered the discussion of *la mentalité primitive* (primitive mentality) in the writings of the French cultural anthropologist Lucien

Lévy-Bruhl (1857–1939), an almost exact contemporary of Husserl's. In 1935, Husserl wrote a letter to Lucien Lévy-Bruhl (Husserl 1935), thanking him for books the anthropologist has sent him. The German text of Husserl's letter was originally transcribed and printed in the appendix to Hermann Leo Van Breda's 1941 doctoral thesis (written in Dutch) and, in this version, was available to Maurice Merleau-Ponty (Van Breda 1992, 156). Jacques Derrida and many others have commented on this letter. In his letter, Husserl shows that he was well read in Levy-Bruhl's writings on anthropology of "primitive" or pre-technological, oral cultures.[3]

Lévy-Bruhl was a prominent French intellectual of the time, a philosopher, sociologist, ethnologist, and theoretical anthropologist, who had a major influence on philosophers such as Ernst Cassirer, psychologists such as Piaget and Jung, as well as anthropologists such as Claude Lévi-Strauss and E. E. Evans-Pritchard. Lévy-Bruhl was particularly interested in the question of whether there is a universal mentality for all humans and whether this mentality goes through stages of development or evolution.

Lévy-Bruhl is best known for his proposal that pre-literate or "primitive" peoples exhibited their own kind of "prelogical" rationality. He claimed that primitives either lived with contradictions or were indifferent to them. He insisted that primitive mentality is different from modern mentality, but rather than being blind to contradictions, as he had earlier put it, he came to see it as having a certain indifference to "incompatibilities" and a "lack of curiosity" about manifest improbabilities, hence allowing room for the mysterious and the mythical. In *Primitive Mythology* (1935), for instance, Lévy-Bruhl points out that, where primitives do recognize contradictions, they reject them "with the same force" as moderns do, however—and this Lévy-Bruhl regards as distinctive of their mentality—there are contradictions that we recognize to which they are insensitive and consequently indifferent (Lévy-Bruhl 1935, xi). Whereas, for example, the European mind assumes an order of causality, the primitive mind ascribes everything to more or less spiritual powers. Primitive thought is essentially "mystical"—there is a felt *participation* and unity with all things; objects are never merely natural, but there is a life-force running through the universe, neither completely material nor completely spiritual, a unifying power running through diverse things (1927, 3).

Husserl was fascinated with Lévy-Bruhl's accounts of the mentalities of pre-literate peoples in Papua New Guinea and Australia (largely based on reports from travelers and missionaries), which seemed to contrast sharply with modern European scientific rationality. According to Lévy-Bruhl, primitive peoples do not experience the natural world in the same way as modern Europeans. Europeans experience nature as ordered and reject entities incompatible with that order (1935, 41). Primitives, on the other hand, experience nature as including what is supernatural. They experience the world holistically, e.g., if one animal is wounded then the whole species feels its pain. "To be is to participate," as he puts it in the *Notebooks*. If a primitive feels unity with a particular totem, then the primitive thinks naturally that he or she is that

totem. Husserl broadly accepts Lévy-Bruhl's account of the worlds of prelit-
erate people enclosed in finite worldviews. Furthermore, Husserl is willing to
acknowledge the plurality and diversity of life-forms, and at the same time he
is careful to avoid the relativism involved in the claim that different life-worlds
are mutually exclusive and mutually incomprehensible. He walked a fine
line: recognizing the factual relativity of everything historical and its related-
ness to its life-world, but also recognizing the *a priori* universal conditions
underlying and governing the experience of worldhood in general.

Husserl thinks that many cultures remain imprisoned in their inherited
world-view without ever questioning it. The ancient Greek theoretical breakthrough, on the other hand, allowed Greek thinkers to discover the relativity
of their worldview in relation to other foreign worldviews (Husserl 1922–
1937, 188). Gradually, a difference emerges between a people's *Weltvorstellung*
(world representation) and what they conceive of as the "world in itself"
(1922–1937, 189). This leads philosophy to a radical "demythification of the
world" (1922–1937, 189) and a stance-taking against traditional values. Here
arises the differentiation between *doxa* and *episteme* (Husserl 1922–1937,
189). With the demythification of experience, we get the rise of "theoretical
experience." Husserl, then, was already writing about the difference between
a historical world and the world of a non-historical people, one enclosed in
myth. Humans living in mythic outlook have a relation to the *Nahwelt* (near
world) (Husserl 1922–1937, 228). Husserl writes in his letter to Levy-Bruhl:

> Naturally, we have long known that every human being has a "world
> representation" (*Weltvorstellung*), that every nation, that every supra-
> national (*übernationale*) cultural grouping lives, so to speak, in a distinct
> world as its own surrounding world (*Umwelt*), and so again every histor-
> ical time in its "world."
>
> (Husserl 1922–1937, 2–3)

Husserl acknowledges that everyone is embedded first and foremost in a
domain of familiarity to which the gives the name *Heimwelt* (homeworld)
(see Waldenfels 1998). To anyone in this familiar homeworld, every other culture appears as a *Fremdwelt* (alien world). First and foremost, one takes one's
orientation from the homeworld, which manifests itself in terms of "familiarity" and "normality." Homeworlds, of course, vary greatly, but each has
its structure of familiarity and strangeness. A community of blind people
will experience blindness as normal. Persons who live on a ship will find its
rocking in the waves to be normal and will find the experience of landing
on *terra firma* to be abnormal. Interestingly, Husserl, in his letter to Lévy-
Bruhl, concedes that "historical relativism proves to be undoubtedly justified
(as an anthropological fact), but also that anthropology, like every positive
science and its universality (*Universitas*), though the first, is not the final word
of knowledge—scientific knowledge" (Husserl 1935, 5). Relativism, then,
has "undisputed justification" as a kind of surface fact that emerges from

comparative anthropological studies. However, he is not content to remain with this apparent relativism. The plurality of historical periods and cultures is not a final fact. Husserl wants to uncover the necessary eidetic laws that govern the very nature of social acculturation and even historicity. As he writes to Lévy-Bruhl:

> For it is in its horizon of consciousness that all social units and the environing worlds relative to them have constructed sense and validity (*Sinn und Geltung*) and, in changing, continue to build them always anew.
>
> (Husserl 1935, 5)

Everything ultimately will be traced back to the *a priori* correlation between intentional intersubjectivity and its horizonal world. This is the "universal *a priori* of history" about which Husserl would write in the famous "The Origin of Geometry" text (Husserl 1936, 371). Husserl concludes his letter with an assertion of the absolute validity of transcendental phenomenology as the final grounding science:

> Transcendental phenomenology is the radical and consistent science of subjectivity, which ultimately constitutes the world in itself. In other words, it is the science that reveals the universal taken-for-grantedness "world and we human beings in the world" to be an obscurity (*Unverständlichkeit*), thus an enigma, a problem, and that makes it scientifically intelligible (*verständlich*) in the solely possible way of radical self-examination.
>
> (Husserl 1935, 5)

Husserl always allows for "relativities" of all kinds, but, in the end, everything has to be traced back to transcendental subjectivity.

Conclusion

Husserl's conception of skepticism and relativism as involving a countersense was attacked in his own day (Kusch 1995), and indeed his own phenomenological claim concerning the *a priori* correlation between forms of givenness and apprehending forms of subjectivity was itself criticized as relativist (by Natorp, Rickert, Cornelius and more recently by Meillassoux (2008), who sees Husserl's correlationism between subjectivity and as not recognizing the possibility of reality in itself outside of all knowability—what Meillassoux calls "the great outdoors"). However, it is indeed part of Husserl's genius to have seen that psychologism, anthropologism, naturalism, and historicism form a complex of philosophical outlooks and tendencies that are remarkably prevalent and yet totally ungrounded in contemporary thinking. He regarded relativism to the very end as a threat to well-grounded scientific knowledge. But the only kind of grounding that Husserl would accept was *transcendental* grounding—the relation of all forms of "sense and being" (*Sinn und Sein*) to absolute subjectivity and intersubjectivity. Ironically, as we have seen, Husserl himself did not escape the charge of relativism levelled at him because of his "correlationism" but he maintained a consistent stance throughout his writings that the relativity of all experiences (whether perceptual, social or historical) is not in itself an argument for relativism. Truth and validity by their nature have an absoluteness and independence that are inherent in their very sense.

Notes

1 Volume One will be indicated 'I,' and Volume Two as '**II**.'
2 For an illuminating history of worldview, see Naugle (2002).
3 Husserl's library, as preserved in the Husserl-Archief Leuven, contains several texts by Lévy-Bruhl: *Die geistige Welt der Primitiven* (1927), the German translation by Margarethe Hamburger of *La Mentalité primitive* (1922), as well as a later edition of that French text (1931); *Le Surnaturel et la nature dans la mentalité primitive* (1931) as well as *La Mythologie primitive. Le Monde mythique des Australiens et des Papous* (1935), the book which is the explicit subject of Husserl's letter, and which also contains the author's dedication.

References

Bambach, C. R. (1995), *Heidegger, Dilthey, and the Crisis of Historicism*, Ithaca, NY: Cornell University Press.

Bolzano, B. (1837), *Theory of Science*, abridged and translated by R. George, Oxford: Blackwell 2014.

Carr, D. (1987), "Phenomenology and Relativism," in Carr, *Interpreting Husserl: Critical and Comparative Studies*, Dordrecht: Nijhoff, 25–44.

Cobb-Stevens, R. (1982), "Hermeneutics without Relativism: Husserl's Theory of Mind," *Research in Phenomenology* 12 (1): 127–148.

—— (1996), "Husserl and the Question of Relativism," *Review of Metaphysics* 50 (1):185–188.

Dilthey, W. (1911), "Brief an Husserl 29.VI.1911," in Edmund Husserl, *Briefwechsel, Teil 6. Philosophenbriefe*, edited by K. and E. Schuhmann, *Husserliana, Dokumente Band III*, Dordrecht: Springer, 1994, 43–47.

—— (1919), *Die Typen der Weltanschauung und ihre Ausbildung in den metaphysischen Systemen, Gesammelte Schriften VIII*, 2nd ed., Stuttgart, Teubner / Göttingen: Vandenhoeck & Ruprecht, 1960.

Heidegger, M. (1927), *Being and Time*, translated by J. Macquarrie and E. Robinson, Oxford: Basil Blackwell, 1962.

Hill, C. O. and G. O. Haddock (2000), *Husserl or Frege? Meaning, Objectivity and Mathematics*, La Salle, Illinois: Open Court.

Husserl, E. (1891), *Philosophy of Arithmetic. Psychological and Logical Investigations*, translated by D. Willard, *Husserl Collected Works X*, Dordrecht: Kluwer, 2003.

—— (1900/1901), *Logical Investigations*, 2 vols., translated by J. N. Findlay, edited by D. Moran, London and New York: Routledge, 2001.

—— (1910–1911), "Philosophy as Rigorous Science," translated by M. Brainard, *The New Yearbook for Phenomenology and Phenomenological Philosophy* 2 (2002): 249–295.

—— (1911–1921), *Aufsätze und Vorträge 1911–1921*, edited by H. R. Sepp and T. Nenon, *Husserliana XXV*, Dordrecht: Kluwer, 1986.

—— (1913), *Ideas: General Introduction to Pure Phenomenology*, translated by W. R. Boyce Gibson, London and New York: Routledge, 2012.

—— (1915), "Brief an Rickert," 20.XII.1915, in *Briefwechsel* vol. 5: *Die Neukantianer*, edited by K. and E. Schuhmann, *Husserliana, Dokumente III*, Dordrecht: Springer, 1994, 177–178.

—— (1919), "Recollections of Franz Brentano," translated by R. Hudson and P. McCormick, in *Husserl Shorter Works,* edited by P. McCormick and F. Elliston, Notre Dame: University of Notre Dame Press, 1981, 342–351.

—— (1922–1937), *Aufsätze und Vorträge 1922–1937*, edited by T. Nenon and H. R. Sepp, *Husserliana XXVII*, Dordrecht: Kluwer, 1989.

—— (1923/24), *Erste Philosophie (1923/24), Erster Teil: Kritische Ideengeschichte*, edited by R. Boehm, *Husserliana VII*, The Hague: Nijhoff, 1965.

—— (1925), *Phenomenological Psychology. Lectures, Summer Semester 1925*, translated by J. Scanlon, The Hague: Nijhoff, 1977.

—— (1927–1931), *Psychological and Transcendental Phenomenology and the Confrontation with Heidegger (1927–31), The Encyclopaedia Britannica Article, The Amsterdam Lectures "Phenomenology and Anthropology" and Husserl's Marginal Note in Being and Time, and Kant on the Problem of Metaphysics.* Ed. Thomas Sheehan and Richard Palmer. Husserl Collected Works VI. Dordrecht: Kluwer Academic Publishers, 1997.

—— (1934–1937), *The Crisis of European Sciences and Transcendental Phenomenology: An Introduction to Phenomenological Philosophy*, translated by D. Carr. Evanston, IL: Northwestern University Press, 1970.

—— (1936), "The Origin of Geometry", in *The Crisis of European Sciences and Transcendental Phenomenology: An Introduction to Phenomenological Philosophy*, translated by D. Carr. Evanston, IL: Northwestern University Press, 1970, 353–378.

—— (1935), "Edmund Husserl's Letter to Lucien Lévy-Bruhl, 11 March 1935," translated by D. Moran and L. Steinacher, *New Yearbook for Phenomenology and Phenomenological Philosophy* VIII (2008), 349–354.

Kusch, M. (1995), *Psychologism: A Case Study in the Sociology of Philosophical Knowledge*, London & New York: Routledge.

Lafont, C. (2000), *Heidegger, Language, and World Disclosure*, translated by G. Harman, New York: Cambridge University Press.

Lévy-Bruhl, L. (1922), *La Mentalité primitive, Bibliothèque de philosophie contemporaine. Travaux de l'Année sociologique*, Paris: Alcan.

—— (1923), *Primitive Mentality*. Translated by Lilian A. Clare, New York: Macmillan.

—— (1927), *Die geistige Welt der Primitiven*, Munich: Verlag Classic Edition, 2010.

—— (1935), *La Mythologie primitive*, Paris: Alcan.

Meillassoux, Q, (2008), *After Finitude: An Essay on the Necessity of Contingency*, Trans. Ray Brassier, London: Continuum.

Mohanty, J. N. (1997), "Husserl on Relativism in the late Manuscripts," in *Husserl in Contemporary Context*, edited by B. Hopkins, Dordrecht: Kluwer, 181–188.

Mohanty, J. N. (1982), *Husserl and Frege.* Indiana: Indiana University Press.

Moran, Dermot (2011), "'Even the Papuan Is a Man and Not a Beast:' Husserl on Universalism and the Relativity of Cultures," *Journal of the History of Philosophy* 49 (4): 463–494.

Moran, D. (2008), "Husserl's Transcendental Philosophy and the Critique of Naturalism," *Continental Philosophy Review* 41 (4): 401–425.

Naugle, D. K. (2002), *Worldview: The History of a Concept*. Grand Rapids, Michigan: William B. Eerdmans.

Soffer, Gail (1991), *Husserl and the Question of Relativism*, Dordrecht: Springer.

Steinbock, A. (1995), *Home and Beyond: Generative Phenomenology After Husserl,* Evanston: Northwestern University Press.

Van Breda, H. L. (1992), "Merleau-Ponty and the Husserl Archives at Louvain," in *M. Merleau Ponty, Text and Dialogues, On Philosophy, Politics, and Culture*, edited by H. Silverman and J. Barry, New Jersey: Humanities Press, 150–161.

Waldenfels, B. (1998), "Homeworld and Alienworld," in *Phenomenology of Interculturality and Life-world*, ed. E. W. Orth and C-F Cheung, Freiburg/München, 72–88.

7 "Open systems" and anti-relativism

Anti-relativist strategies in psychological discourses around 1900

Paul Ziche

Protagorean relativism and the general standpoint

Who wants to be a relativist? The term "relativism" seems to be a concept characteristic for the period around 1900—but it typically is not used as a concept for affirmative self-description (Köhnke 1997). Quite the same holds for "psychologism," and many authors (Husserl is a prominent example) establish close connections between relativism and psychologism. Relativism is typically seen as posing a threat: it questions the very possibility to arrive at, justify, and consistently hold absolute values—logical, epistemological, ethical, aesthetic values—and thus potentially affects all fields of philosophy and the sciences. This can be captured in a radicalization of the "man is the measure of all things" phrase: for the relativist, *individual* man becomes the measure of all things, or, in a widely used rephrasing, each and every *standpoint* becomes equally acceptable because any clear method is lacking that might allow it to come to a decision among these standpoints.

A prominent statement of this generalized Protagoreanism is given by Hugo Münsterberg (1863–1916) in his ambitious book on *Eternal Values* from 1909. Münsterberg, active as a psychologist and philosopher and a key figure for establishing the discipline of psychology in the United States,[1] sees his own time as being dominated by a relativism in which all values come to depend "upon our special standpoint." While we, as practicing scientists, as religious believers, as artists or social reformers, do believe that there are absolute values, "all these convictions and beliefs, these faiths and inspirations, must fade away, it seems, as soon as the philosopher begins to examine them"—and he summarizes this in a very concise characterization of relativism:

> Everything is relative; everything is good only for a certain purpose, for a certain time, for a certain social group, for a certain individual. ... Philosophical skepticism and relativism are thus the last word, and their answer harmonizes with a thousand disorganizing tendencies of our time.
>
> (Münsterberg 1909a, 1–2)

Münsterberg makes clear from the outset that he intends to reject this relativism. In his own anti-relativist argumentation, he identifies two main opponents: a philosophy that recognizes a merely "pragmatic value" in goodness, beauty, progress, truth, peace, and religion, i.e., in everything that had been thought of being absolutely valuable; and a psychological approach that claims that what "we dream of eternal values should simply be explained psychologically like the fancies of a fairy-tale" (Münsterberg 1909a, 2).

His strategy in refuting relativism is highly interesting. He wants to promote an "idealism," a Kant- and Fichte-inspired account of the necessity of absolute values that interprets "reality in voluntaristic spirit" and acknowledges "the fundamental character of the purposive activity" (Münsterberg 1909b, 334)—another term that Münsterberg uses to denote the fundamental role of the will is "teleological." The idea is that a naturalistic reduction of mental processes becomes impossible as soon as the will is established as a fundamental fact. His theory

> will show that idealism is justified, nay, is demanded, by true science and true philosophy, that the believers are right and the pragmatists wrong, and that we may stand firmly with both the feet on the rock of facts, and may yet hold to the absolute values as eternally belonging to the structure of the world.
>
> (Münsterberg 1909a, 2)

While Münsterberg presents his position in strongly dualist terms, and while he criticizes psychology for reducing eternal values to fictitious ideas, he nevertheless asks us to stick close to "the rock of facts," and this includes giving a role to psychological research. For Münsterberg, being anti-relativist (and anti-psychologist) does not imply that he also takes a strictly anti-psychological attitude. The key step here consists in changing our conception of what Münsterberg calls the "structure of the world." As with many other value theorists, values are not seen as an added layer that is superimposed upon a body of neutral facts. Rather, values themselves become intrinsic properties of states in the world.

The very term "structure of the world" indicates that Münsterberg is aware that this changes our understanding of what "facts" or "world" might mean. Münsterberg's strategy can be described in terms of a broadening of our understanding of "facts" and of the "world" that leads to what we may call a *general standpoint* that is clearly distinguished from just accumulating a number of neutral facts. He does not opt for a response to relativism in terms of selecting one standpoint and preferring this standpoint above others. This would lead into serious difficulties of the kind that in more recent epistemology are discussed under the title of "no meta-justification": Singling out a particular standpoint would require justificatory arguments that can be accepted among the competing standpoints, and that is just what the coexistence of rival standpoints denies. Rather, he intends to arrive at a point at which we are no longer forced to take a decision between rival options, but that still can be called a standpoint, for instance, in virtue of it being itself a fact.

This strategy is closely related to other forms of generalization in philosophy that are prominent in the period around 1900:[2] Husserlian phenomenology can be described in similar terms, as can a hermeneutic approach that denies the independent existence of individual standpoints. What is particularly interesting about Münsterberg's version is that he wants to steer a *middle course* that allows him to maintain absolute values *and* to incorporate results from psychology at the same time.[3] This chapter looks more closely into the role of psychology and of philosophical conceptualizations of psychological theories within debates concerning relativism around 1900. Its key strategy is to show how the turn toward higher forms of generality is prominently present in psychological and psychology-related discourses in this period and how this could support hopes for a non-reductionist way of including the results and practices of the special sciences into larger foundational projects.

Toward a general theory of values: "Bring values home"

The status of values becomes itself the topic of a novel subfield in philosophy.[4] Discussions in the philosophy of values are typically characterized by strong dualisms, in particular the dualism between the claim that values can be, and need to be, *absolute*, as opposed to the *relativist* attitudes that became associated with a psychological or naturalistic analysis of values. This dualism finds its clearest expression, at least in the English-speaking world, in the discussion between Münsterberg, who argues for the necessity to arrive at absolute or "eternal" values, and Wilbur Marshall Urban (1873–1952), who argues, against Münsterberg, for the importance of a psychological analysis in the theory of values.[5]

Münsterberg stages this debate explicitly in dualist terms and shows that the dualisms in the philosophy of values can be related to a number of other conflicts in the philosophy of his time: the positivist vs. the idealist, the relativist vs. the absolutist, a psychological vs. an epistemological stance, the analysis and explanation of facts vs. a teleological system (1909b, 329). Münsterberg himself argues for a revival of the standards of Kantian critical philosophy in order to counteract the relativism of his days: "We simply must once more force on our present-day relativism the fundamental categories of critical philosophy: we must go back to Kant and from Kant to Fichte"—or, with the Neo-Kantian outcry: "Back to Kant!" (1909b, 337–388)[6]

Even though Münsterberg includes a "psychological" approach in his list of dualisms, he nevertheless emphasizes that his anti-relativist attitude can be seen as a continuation of psychological approaches. He has three arguments to this effect. Firstly, the idealist in Münsterberg's sense can "accept ... and appreciate ... the psycho-sociological studies of the relativist" (1909b, 329), as long as they are not presented as the ultimate principle in a study of values.

The relativist, on the other hand, cannot include the absolutist stance into her theories because she cannot accept absolute values. Secondly, the prevalent atomism in psychology—the analysis of mental phenomena into their ultimate elements, forming the methodological foundation of associationist psychology—can pave the way toward an idealist understanding of these phenomena: "Consistent psychological atomism of the will and radical teleological philosophy of the will belong most intimately together" (1909b, 332). The idea seems to be that the importance of the will and the irreducibility of mental phenomena to these elements can only be understood if we first boil down the mental phenomena to their ultimate elements. The third argument points out that eternal values need to be represented in consciousness in order to become efficacious as principles that govern our activities.[7] The changing ontology of (mental) states that has already been adumbrated in section 1 can summarize these arguments: we need to accept that the "givenness of the objects and their existence involves valuation" (1909b, 333). This can lead to a refinement of the anti-naturalist stance with respect to psychology. Münsterberg criticizes the atomism that he, just as many of his contemporaries, detects in the more empiricist, reduction-supporting strands of psychology as being one-sided; associationist psychology in its elementarist approach neglects the complexity of states that his teleological argument required, and that he thought to be able to discover via psychological experience: "The standpoint of the naturalist is an artificial one; it involves certain abstractions" that distort the complex states under consideration (1909a, 13).

The exchange between Münsterberg and Urban starts off in dichotomically dualist terms. This makes it the more surprising that there is an unexpected and important harmony with respect to the role they ascribe to psychology. Urban's goal is stated, explicitly, in terms of unification and of a scientific attitude; he strongly opposes what he calls "misologistic" tendencies (i.e., anti-intellectualist or irrationalist tendencies in psychology/philosophy; Urban 1909, VIII) as well as the Kantian distinction between the empirical and the *a priori* and aims at pursuing and further developing science to its fullest extent. In the relevant passages, he refers to, among others, Alexius Meinong (1853–1920) and Christian von Ehrenfels (1859–1932) as authors who pursue a related agenda in unifying the empirical and the *a priori* in the theory of values:

> The desideratum, therefore, seems to be to find a method which shall unite in some more fruitful way the descriptive and the normative points of view, a method which shall know how to interpret the norms of the so-called "intelligible" will in terms of the laws of the "empirical" will.
>
> (Urban 1909, 6)

One way to achieve this unification is precisely by emphasizing the ubiquity of psychology, in a psychologistically sounding phrase that at first sight lends support to Münsterberg's charging Urban's value theory as being psychological: "Strictly speaking there is no problem, scientific or unscientific, which does not have its psychological side. Not only the questions, but also the objects in connection with which these questions arise, belong in the first place to the psychical life" (Urban 1909, 9). We can now see why this need not be incompatible with Münsterberg's account: if complex states can be (psychologically) given to us, the psychological character of these states can indeed support a unification of the descriptive and the normative.

We get into a rather surprising situation. The two opponents in a debate that was presented in dualist terms share profound convictions. It did not go unnoticed that the dualist opposition between Urban and Münsterberg did not seem to be adequate to their greater ambitions. A particularly salient case is presented in a 1917 paper on the "Theory of Values" by Columbia philosopher Herbert Schneider (1892–1984) in the *Journal of Philosophy Psychology and Scientific Methods*, of which Schneider was the editor from 1924–1961—the precursor of today's *Journal of Philosophy*. Schneider, too, argues for a more positive attitude toward psychology in the theory of values, that would take away the "odium of the 'merely' and the 'nothing but'" that "psychology and empiricism have to bear" (1917, 144). His main argument consists in reconstructing values as *complex* three-place predicates that combine a valuable object, an organism or activity to which it is valuable, and an end or purpose for which it is valuable (1917, 146). Neglecting one (or more) of these factors by way of abstraction is as inadequate as the reaction on the side of "absolute idealists and realists" who cry, "'relativism and subjectivism!'" (Schneider 1917, 147). Leaving the reference of values to specific ends out of consideration, making them "irrelative," would only render these impoverished values, in a pragmaticist argumentation, "irrelevant." Only if we respect the complexity of value predicates can we study them in their natural habitat:

> The present need is that psychology study values at home, in their natural and specific situations. They can not rightly be studied as abstractions; they must be studied in their functional relationships, and this involves a study of all three factors of the value situation in their proper and specific relations.
>
> (Schneider 1917, 148)

As a consequence, Schneider arrives at what looks like a relativist position; values, according to his three-dimensional analysis, are indeed relative in the sense that they are "not an absolute unchanging piece of reality, but a characteristic of nature by means of which organic activity is made possible and carried to its perfection" (1917, 154). However, this does not imply that a theory of absolute values needs to be or can be opposed to his understanding of values as specific and concrete—values only function in acts of valuation and cannot be studied independently of these complex processes. There simply is no theory of values in the abstract; emphasizing the complexity of evaluative states does not so much relativize values but rather gets us closer to understanding their nature.

The idea that psychology may be used for arriving at a philosophically sensible, but nevertheless non-naturalist theory of values was shared more broadly. One prominent example: one of the key authors both for value theory and for innovations in psychology, Christian von Ehrenfels, in his 1893 treatise on *Werttheorie und Ethik* (*Value Theory and Ethics*) contrasts the trend toward the naturalistic "objectivation" of values with a "psychological" analysis. While the naturalist overstresses the role of the intellect, the psychologist can appreciate the psychological capacity "that alone is capable of creating values" (von Ehrenfels 1893, 87), namely the "emotional dispositions" that are at work in all questions concerning values. It is precisely the dimension of feelings, as studied by psychology, combined with an integrative, non-abstractive account of the human mind, that again is—as all the authors presented so far stress—not only not at odds with, but even deeply engrained in, the spirit of psychology, that is required for a non-naturalist theory of values.

Relationalist anti-relativism in psychology

Where does psychology deal explicitly with relativism? One line of argument has already been presented in reference to the broadly shared anti-atomistic attitude, with its anti-empiricist and anti-naturalist implications. Another line can be found in the most quantitative sub-field of psychology, namely in psychophysics. The key achievement of a mathematized psychophysics, the Weber-Fechner-law, is standardly introduced in the psychology textbooks of the period in terms of the *relativity* of sensations.[8] One example: Theodor Ziehen (1862–1950), clinical psychologist, philosopher, author of widely read handbooks, presents Wilhelm Wundt as considering the Weber-Fechner-law as just a special case of the "general law of the relativity of our mental processes in general" (Ziehen 1924, 65), namely the principle that there is no absolute but only a relative measure for their intensity.[9] It is in this way that *Wundt* treats Weber's law in his own *Physiologische Psychologie* (1893, 393). Weber's law, according to Wundt, does not refer to sensations themselves but to apperception as the operation of relating sensations to one another; it is a law not for states but for *relations* and for relative measurements. Ziehen generalizes these ideas into a philosophical position that he calls *Relationismus,* which is based upon the idea that "the things studied in natural science are, in their essence, nothing but relations" (1927, 23).

It is obvious that this program is related to other philosophical and psychological programs that, in many respects strongly influenced by *Gestalt* theory, introduce novel types of objects that are claimed to be more general than traditional object categories would us have it, and that thus go beyond traditional sub-divisions of the field of philosophy. Alexius Meinong is a key protagonist in these debates, with two strong claims that become intimately related in his texts: The claim that he succeeded in initiating an innovatively general theory of values (Meinong, 1911, 132), and the claim—characteristic for his *Gegenstandstheorie*—that we need to adopt more general object categories. His brief text on *Für die Psychologie und gegen den Psychologismus in der allgemeinen Werttheorie* from 1911 highlights these claims. Meinong, too, intends to reject the charge of psychologism but nevertheless wants to continue making use of psychology. Psychology even remains of key importance because our attitude with respect to values cannot be understood as a purely intellectual endeavor; rather, our emotional life, the *Gemütsleben*, plays a key role here, and consequently psychology should not remain restricted to analyzing human mental life as a logical process.

The innovative character of his theory is mirrored in Meinong's terminology that one can hardly see other than as coining consciously unwieldy neologisms. Examples are terms referring to the role of feeling in cognitive processes: *Urteilsgefühle* and *Wissensgefühle,* feelings of judgment and feelings of knowing (1911, 136),[10] *Vorstellungsgefühle, Urteilsinhaltsgefühle*, feelings of representations and of the content of judgments. Other terms seem to be closer to existentialist sentiments but still refer to feelings that are involved in judgments about objects: The feelings of *Daseinsleid* and *Daseinsfreude*, the grief or pleasure related to the existence of an object (1911, 136). What these terms already suggest, with their combination of subjectively accessible emotion terms and epistemological concepts, underlies Meinong's argument in favor of psychology and at the same time against a psychologism: psychology itself needs to concede, according to Meinong, that "pre-theoretical" accounts of values need to be included in any psychological theory of valuation. Meinong makes the strong ontological implications of these ideas explicit by requiring that any analysis of value judgments has sufficient ontological "latitude" (1911, 139): feelings need not be initiated by actual encounters with objects, but also potential objects, or objects that I only potentially might own, can be valuable for me. What we need is a combination of a theory of super-personal, absolute values with the empirically well-supported personal and relative character of values that psychology has tended to neglect (1911, 141). Meinong refers to a large number of allied authors, operating between philosophy and psychology, and all building forth upon an idea that he ascribes to Wilhelm Windelband,[11] namely, the idea that "the close relationship between intellectual and emotional experiences has been unduly neglected by the, otherwise quite adequate, psychological tradition" (1911, 141).

A key topic among these authors concerns the program of unification in psychology and includes the integration of cognitive/logic-related and emotional states. A characteristic example of this strategy can be found in a 1905 paper on the cognitive value of aesthetic judgments by Edith Landmann-Kalischer (1877–1951), one of the authors on Meinong's list. Landmann-Kalischer, best known for her involvement with the George circle, departs from the "view, generally shared in science" that values, in particular aesthetic values, are "subjectively determined everywhere" and that value judgments are not aimed at cognition (1905, 264). But she sees it as timely, urgent, and possible to remove the odium of subjectivity from aesthetic valuations, and intends to upgrade feelings by understanding them in analogy with sense perception, i.e., with the paradigm of objectivity-related judgments. In the concluding passages of her paper, she strongly emphasizes that this indeed means a strong revision of earlier attitudes in philosophy:

> While in earlier times [the times of Locke] the subjectivity of sense impressions was demonstrated by showing that they do not differ from feelings, we, today, have to establish the objectivity of the properties in things grasped via feeling by placing them in one line with sensory qualities.
>
> (1905, 328)

Feelings, thus, should be understood in such a way that they can claim objective status. This inverts our epistemological expectations regarding subjectivity and objectivity, and regarding the distribution of these epistemic characteristics among feelings and sense perceptions in that feelings, and subjectively determined values, are viewed as candidates for objectivity: While it is, according to Landmann-Kalischer, undeniable that values have subjective conditions (1905, 267), it remains possible that, while "acknowledging fully this being subjectively conditioned" of values, values still can be valid in a non-subjective, in a subjectivity-transcending way—and that in this sense value judgments are analogous to judgments about sensory qualities and are geared toward cognition (1905, 267–268).

Landmann-Kalischer has two strategies to offer in support of this view. In both cases, she introduces object types of a higher level of generality with the intention that these higher levels of generality objectify what is subjectively conditioned: *Gestalt* qualities on the one hand (e.g., 1905, 279),[12] relational structures on the other are introduced to this purpose (1905, e.g., 274). Landmann-Kalischer's paper itself attracts a wider audience and is embedded into contexts in which precisely these operations of generalization occupy center stage.[13]

We see a number of larger issues at work here. If feelings, as the traditional paradigm for subjective mental states, can be objectified (and if, as a consequence, aesthetic judgments acquire cognitive status), this means at the very least that an emotional component in one's theory is not sufficient for an indictment as being relativist. Put more generally: the strategy seems to be one of redrawing typical boundary lines in the field of philosophy (here that between objectivity and subjectivity) in order not to directly refute the relativist but to fruitfully accommodate precisely the key tenets of the relativist into a yet broader picture that allows a return to traditional philosophical values such as scientificity. The resulting picture is one that needs to, on the one hand, incorporate the openness and indeterminacy characteristic of the relativist and, on the other hand, has to introduce high-level concepts that can govern the totality of open and indeterminate ideas in a way that can be seen as scientific. Typically, *Gestalt* properties and relational structures are deemed ideal candidates for filling in this agenda.

In all cases, this picture has strongly anti-dualist implications. The simple dualism between subjective and objective states does not work: "I am of the opinion that we should become suspicious with respect to such purely subjective elements of consciousness" (Landmann-Kalischer 1905, 273); "subjectivity and objectivity do not reside in particular phenomena in/of consciousness (it is not the case, for instance, that any feeling is subjective while any mathematical thought is objective)," but rather derive from the "constellation, the law-governed dependencies in which we encounter a phenomenon" (1905, 276). The stable relational linking of an object to our mental states is viewed as being a property of this very object, thereby extending the notion of properties so as to include also relational properties (1905, 274). *Gestalt* properties that are declared by Landmann-Kalischer to be directly perceivable, despite their being conceptually complex (1905, 279), again provide the best examples. With respect to value judgments, this relationalist account of values implies that values cannot be reconstructed in a (reductively) naturalist fashion.[14]

One of the most striking features of both Meinong's and Landmann-Kalischer's accounts, as well as of the other relationalist accounts, is that what seems to be the most individualist and subjective, namely emotions and feelings, becomes incorporated into an argument that is directed at new levels of generality, and at maintaining high standards of scientificity and at keeping in contact with the special sciences throughout.[15] At this point, it becomes possible to return to debates in the philosophy of value and to ask how the ideal of strictly scientific openness becomes operationalized in these debates.

Relativism, pragmatism, phenomenalism, correlativism, realism …: Integrative accounts in innovative forms of philosophy

We thus arrive at a highly distinctive trend that pervades philosophical and psychological discourse around 1900: an agenda of aiming at large-scale integrations that reach beyond the traditionally established demarcation lines in philosophical discourse can be found in (at least at first sight) highly diverse contexts. In a number of cases, these integrative accounts lead to newly labeled forms of philosophy. Some of these accounts, in their reaction against relativism, will be sampled in this section.

Examples for the prominence of strongly relationalist programs can be multiplied easily. Ernst Cassirer's favoring of functional concepts above substance concepts is a prominent example, as is Max Frischeisen-Köhler's (1878–1923) discussion of the problem of reality.[16] Maximilian Beck (1887–1950), who obtained his Ph.D. in Munich with Munich phenomenologist Alexander Pfänder before emigrating to the United States, even creates an entirely new label, *Korrelativismus* (Beck 1928), for a relationalist philosophy that intends to overcome the disjunctions between subjectivism and objectivism, idealism and realism, and to support a concrete form of interaction between man and the world of objects. (There are Heideggerian undertones here, and Beck also refers to Spengler.)

Texts by William James (1842–1910) and Oswald Külpe (1862–1915) can illustrate how the notion of relativism and the problem of abstraction migrate between philosophy and experimental psychology. William James devotes an entire chapter in *The Meaning of Truth* to a discussion of "Abstractionism and 'Relativismus'," clearly referring to a German-language discourse and including an intense discussion with Münsterberg's account of "eternal values" (James 1909). He inverts Münsterberg's criticism that pragmatism is an unjustified form of abstraction and strongly rejects what he labels as "vicious abstractionism," and he also sees this vicious attitude at work when authors such as Heinrich Rickert (1863–1936) or Münsterberg charge pragmatism for being relativist (James 1909, 263).[17] While conceiving of a concrete situation abstractively by singling out some salient feature is a useful strategy, things go wrong when "we proceed to use our concept privatively; reducing the originally rich phenomenon to the naked suggestions of that name abstractly taken, treating it as a case of 'nothing but' that concept" (1909, 249): "Abstraction, functioning in this way, becomes a means of arrest far more than a means of advance in thought. It mutilates things." This analysis of the dangers of an abstractionist account serve James to characterize pragmatism in terms of adding "concreteness" to our way of analyzing concepts (1909, 262–263). The concept of abstraction plays a double role in his argument. James reacts against Rickert's and Münsterberg's criticism that pragmatism is unable to conceive of the abstract notion of truth, and turns the table by charging his critics with themselves operating on a level of abstraction that makes it impossible "to give any account of what the words may mean" (1909, 266) or that abstracts the relevant notions—such as "truth"—"from the universe of life" (1909, 268). The discourse on relativism in Germany, according to James, suffers from precisely such an over-abstraction (1909, 270–271).

The notion of abstraction, and the critical rejection of abstraction as leading away from concrete reality, find their way into experimental practice in psychology and in philosophical theorizing that is strongly influenced by psychological experiment. Oswald Külpe, a Wundt-pupil, later emancipating himself from Wundt and becoming the central figure of the "Würzburg school of experimental thought psychology,"[18] is the most visible protagonist for mutually incorporating psychology into philosophy, and vice versa, in a way that can be analyzed in terms of an integrative account of abstraction. At this point, we can start looking into psychological practice. In his main work in philosophy, *Die Realisierung*, Külpe emphasizes that abstraction does not imply that we move away from the realm of things; abstraction is not an *Entdinglichung*, a de-reification, and it does not imply a loss of individuality, an *Entindividualisierung* (1912, 132–133). Concrete and abstract objects are not fundamentally different (1912, 137); concrete objects, as a limit concept in cognition, also have abstract characteristics.

Evidence for these claims can be found, Külpe claims, in psychological experiments, where Külpe explicitly endorses introspection as a method in experimentation. The evidence is, in fact, multiple: In his experiments, Külpe regularly encounters states that are, according to traditional criteria, abstract, devoid of imagistic content, but that can nevertheless be experienced as distinctive and complex inner states. Another line of argument derives from Külpe's experiments on "subjectivation" and "objectivation," that is, his study of those factors that determine whether we take a particular mental representation as being subjective or objective. Again, his experiments show that an experience's being deemed subjective or objective is not due to an inherent feature of a representation but depends upon other factors (such as the particular task that is studied in an experiment) (Külpe 1912).[19] Külpe thus studies, in psychological experiment *and* philosophical theory, what James asked for in his plea for concreteness. What is most remarkable about his research is his confidence in indeed being able to apply the experimental method, in continuity with the experimental practice of the special sciences (this claim, of course, would require closer scrutiny), in order to gain deeper insight into philosophical concepts and theories. As in virtually all authors presented so far, it is the anti-elementarist, anti-associationist, anti-imagistic (in the sense of: against a notion of ideas that understands them as copies of sense impressions) results of his experimental research that allow him to confidently claim that his brand of psychology does not fall prey to charges of relativism or psychologism.

Summary: "Open systems" in scientific philosophy and psychology

Reacting against the challenge of relativism means taking a position in a situation where there are too many options available; too many standpoints without a basis for choosing among them. What has been presented in a caleidoscopic overview in this chapter points at a rather surprising, but also widely adopted, strategy. Both Münsterberg and, even more explicitly, Rickert phrase this strategy in terms of opening up the rigorous notion of "system." Münsterberg asks whether the philosopher cannot and should not "find in his own system fullest room for the free unfolding of the relativistic knowledge" (1909b, 329), and Rickert poses the question: "Couldn't we find our strength in consciously renouncing closure?" In discussing the "system of values," Rickert finds a compact phrase for the openness that is required for incorporating a broad range of positions or attitudes, and for accounting for the development and progress of science: he argues for "*open* systems" (1913, 297). His motivation for maintaining a place for systems in philosophy derives from a strongly anti-systematic trend that he perceives in the philosophy and in the broader culture of his time, and that he relates to the influence of Friedrich Nietzsche's thinking. Rickert concedes that the very term "open systems" is, on the surface, contradictory (1913, 297). His reaction toward this contradiction consists in specifying more precisely what open systems need to do: he requires openness for the more specific purpose to account for the *Unabgeschlossenheit,* the non-closedness of history, and the factors that are used in systematizing history can "reach beyond history without conflicting with it" (1913, 297). Openness, refraining from aiming at achieving closure, can be viewed as "strength" (1913, 298): this, then, requires novel concepts for systematicity that are sufficiently open so as to include (in Rickert's case: historical) change; "order" is Rickert's example for such a concept that, in itself, does not impose any kind of rigorously structured hierarchy.[20] More specifically, Rickert argues for a co-existence between connected systems of values that transcend historicity, and the concretely determined events in history (1913, 300), in what can be read as a rather direct rendering of the idea of a hermeneutic circle.

There are two main lines for further analyzing this strategy of thinking in terms of "open systems." On the one hand, one can further contextualize this strategy by relating it to broader trends that we can detect in a number of debates around 1900: in various fields, we find a move toward increasingly higher levels of *generality* that are counterbalanced by a critique of abstraction. One of the most characteristic achievements that are to be gained by stepping up to these higher levels is that what previously appeared to be different concepts and ideas become *harmonized* once this higher level is reached. Since this move can be detected in areas that are paradigms of scientificity[21] as well as in broader philosophical and world-view movements, this strategy sheds a novel light on the role of relativism. If divergent ideas can coexist, by becoming abstractions or specifications within a more general framework that is thoroughly scientific, relativism need no longer be seen as a stance that departs from, or requires, a loosening of our standards; rather the opposite becomes possible: as a consequence, apparently divergent or opposed notions such as subjectivity and objectivity, abstraction and concreteness, generality and experience can become integrated in these debates. Within a framework of "open systems," relativism ceases to be a threat—rather, it becomes a foundational ingredient in science-inspired and science-directed thinking. Why this openness got lost, some way between the period discussed here and more recent philosophical attitudes, thus becomes a pressing, but by no means clearly resolved, problem.

This has important implications for our understanding of what it means to be scientific, and more particular of the status of holist movements in the period around 1900. Anne Harrington has analyzed "holist" movements in psychology and science as deeply ambivalent events, torn between a drive toward re-enchanting our world-views and the scientific ambitions of the protagonists (Harrington, 1996). Where this chapter takes issue with Harrington's approach is that Harrington rather strongly emphasizes "irrational" elements in "holist" ideas (e.g., Harrington 1996, 27, on Külpe)—rather, it seems to be important for the protagonists in this debate that they present their ideas as being scientific, even as scientific in a particularly rigorous sense.[22]

More systematically, it is tempting to relate these "open systems" (and related notions) to a discussion of relativism in terms of two-level accounts. Martin Kusch (2017) expresses sympathy with epistemic relativism, and in order to do so, "the relativist must formulate his position in a way that involves two perspectives" (2017, 4692), namely a first-order perspective that "we happen to have because of contingent historical circumstances," the epistemic system that we happen to have adopted. Combined with that, we need a second perspective that is based in a reflection "on the contingency of one's epistemic practices and standards" (2017, 4693). In this picture, a "relativist second-order perspective" can be held together with being committed to a particular first-order perspective because "the second-order perspective does not have epistemic principles that directly compete with those of the first-order perspective" (2017, 4694).

The anti-relativist strategies presented in this chapter can indeed be seen as analyzing the problem of relativism and the adequate response in terms of two different perspectives, where one perspective guarantees the kind of openness and individuality that was thought of as leading to relativism while the other integrates these many standpoints into an ordered whole. In the cases of Rickert and Münsterberg, we have the open horizon of historical change that still can be ordered within open systems, and the infinity of personalized value attributions that presuppose a framework of absolute values. Compared to the arguments in Kusch's article, however, the perspectives are inverted. For Kusch, it is the reflective stance of the second-order perspective that allows for a rational form of relativism, while both Münsterberg and Rickert (and many others, as indicated in this chapter) argue for higher-level principles that point *beyond* relativism.[23]

In the anti-relativist strategies that have been presented here, it is on the higher level that the real commitment is asked for. We find various ways of justifying this commitment: In many cases, it is particular argument forms that are used on this higher level (Münsterberg, for example, makes ample use of transcendental arguments in his theory of absolute values); but just as important is the integrative or unificatory function of this higher level. More argument is needed in order to elaborate precisely how this higher level is thought to cooperate with the lower levels. This gets further complicated by the strong continuities among the levels. Take the example of feelings, which are typical candidates for generalizing moves that are intended to employ the characteristics of feelings (such as their concreteness, their phenomenological richness, there directness) also on the higher level. Another phrase that may capture this difficulty is: what the anti-relativist authors typically search for, are novel forms of systematization and, via a systematic ordering, also of justification that, nevertheless, keep close contact with traditional notions and arguments. Feelings, again, provide an illustrative example. It is remarkable to see the extent to which these theories dare to move beyond established boundaries between philosophical theories, between philosophy and the special sciences, and between novel and traditional philosophical concepts. Looked upon in this way, the anti-relativist arguments as presented here, and the role of psychology in these arguments, can also serve as a basis for further refining the status of philosophical projects such as Husserlian phenomenology and (Diltheyan and others') hermeneutics.

Acknowledgments

I am very grateful to the editors for critical comments on an earlier version of this text. All translations by the author.

Notes

1 Münsterberg bridges the gaps between European and American philosophy and psychology on a number of levels: a pupil of Wilhelm Wundt, he was invited to the U.S. by William James and later institutionalized academic exchange between the U.S. and Germany. One of his most visible tasks was the organization of the 1904 "Congress of Arts and Sciences" in St. Louis, for which he projected a comprehensive system of the sciences (on this topic, see Ziche 2008).
2 On the prominence of the notion of generality in science-related discourses around 1900, see Hagner and Laubichler (2006).
3 In a next step, it clearly would be relevant to compare the authors and ideas presented here in more detail to the philosophico-psychological projects of, in particular, Husserl and Dilthey.
4 For a contemporary overview, see Messer (1926).
5 Urban, a professor at Dartmouth and Yale, was strongly influenced by ideas form Ernst Cassirer, who also succeeded him at Yale.
6 As is frequently done in the texts discussed here, Münsterberg also sees a dualist opposition between a systematic spirit in philosophy on the one hand, and the "impressionistic philosophizing" current in his time on the other (1909b, 335).
7 See, e.g., Münsterberg (1909a, 48). This passage is significant: Münsterberg starts from the broadly Kantian conviction that all our mental states remain dependent upon the conditions of consciousness; in a second step, he reads this transcendental standpoint as implying that even the world of absolute values does not exist "eternally separated from our consciousness."
8 Meinong discusses Weber's law extensively in a number of contexts; see, e.g., Meinong (1896); Meinong (1888).
9 On the Weber-Fechner-law, see also Ziehen (1913, 467), again in terms of the "general relativity of psychical processes."
10 On the notion of "*Gefühlsgewißheit*" and on "*Wahrheitsgefühle*," see Albrecht (2015); Ziche (2015).
11 On Windelband and relativism, see Kinzel (2017).
12 The most comprehensive discussion of *Gestalt* psychology still is Ash (1995).
13 Landmann-Kalischer's texts are referred to frequently in a number of the texts discussed here, e.g., in Urban's text that is discussed in section 3; see also Urban (1907). On Landmann-Kalischer, see, e.g., Reicher (2016).

14 Landmann-Kalischer's list of authors who recognize the *"Sonderstellung des Wertes,"* the "special position of value," include, among others, Simmel, Meinong, Jonas Cohn, and in particular Ehrenfels (Landmann-Kalischer 1905, 264–265).

15 In this chapter, "science" is always used in the broad sense, thus including the natural sciences, the humanities, and the social sciences under this term.

16 Frischeisen-Köhler, philosopher and educationist at Halle, strongly influenced by Dilthey, carries on with the relationalist programme, and he generalizes this program yet further in the sense that we not only need to focus upon relations everywhere but also must turn toward the "universal relational connection," *dem universellen Beziehungszusammenhang* (Frischeisen-Köhler 1912, 12–13). His references include authors from classical hermeneutics and German idealism: Goethe, Schelling, Schleiermacher, Jacobi, Bergson, Dilthey (Frischeisen-Köhler 1912, 48–50).

17 For a discussion of pragmatism as a form of "generalization," see Perry (1907).

18 On the "Würzburg school," see Kusch (1999) and Ziche (1999).

19 The same kind of argument is, again, also pursued in Külpe's experiments concerning "abstraction" (Külpe, 1904).

20 On theories of order around 1900, see Ziche (2016).

21 Mathematics is the most prominent example; see Ziche (2008, ch. 6).

22 In the light of the prominence of integrative analyses of value systems, and of the important role of feelings in evaluative, yet scientific contexts, Daston and Galison's (2007) account of objectivity in terms of disinterestedness and depersonalization also needs to be critically questioned.

23 Rickert, in his 1913 paper on the system of values, makes clear that being a system, in his view, requires a "principle," and more specifically a principle of completeness that goes beyond the mere juxtaposition of (historically determined) facts or ideas (Rickert 1913, 298).

References

Albrecht, A. (2015), " 'Wahrheitsgefühle'. Zur Konstitution, Funktion und Kritik 'epistemischer Gefühle' und Intuitionen bei Leonard Nelson," in *Ethos und Pathos der Geisteswissenschaften. Konfigurationen der wissenschaftlichen Persona seit 1750,* edited by R. Klausnitzer, C. Spoerhase and D. Werle, Berlin: de Gruyter, 191–213.

Ash, M. (1995), *Gestalt Psychology in German Culture, 1890–1967: Holism and the Quest for Objectivity,* Cambridge: Cambridge University Press.

Beck, M. (1928), "Die neue Problemlage der Erkenntnistheorie," *Deutsche Vierteljahrsschrift für Literaturwissenschaft und Geistesgeschichte* 6: 611–639.

Daston, L. and P. Galison (2007), *Objectivity,* New York: Zone Books.

Ehrenfels, C. von (1893), "Werttheorie und Ethik," *Vierteljahrsschrift für wissenschaftliche Philosophie* 17: 76–110, 200–266, 321–363, 413–425; 18, 22–97.

Frischeisen-Köhler, M. (1912), *Das Realitätsproblem,* Berlin: Reuther & Reichard.

Hagner, M. and M. Laubichler (eds.) (2006), *Der Hochsitz des Wissens. Das Allgemeine als wissenschaftlicher Wert,* Zürich and Berlin: Diaphanes.

Harrington, A. (1996), *Re-Enchanted Science: Holism in German Culture from Wilhelm II. to Hitler,* Princeton: Princeton University Press.

James, W. (1909), "Abstractionism and 'Relativismus'," in James, *The Meaning of Truth. A Sequel to Pragmatism,* Ann Arbor: University of Michigan Press, 1970, 246–271.

Kinzel, K. (2017), "Wilhelm Windelband and the Problem of Relativism," *British Journal for the History of Philosophy* 25: 84–107.

Kleinpeter, H. (1905), *Die Erkenntnistheorie der Naturforschung der Gegenwart. Unter Zugrundelegung der Anschauungen von Mach, Stallo, Clifford, Kirchhoff, Hertz, Pearson und Ostwald*, Leipzig: Johann Ambrosius Barth.

—— (1912), "Zur Begriffsbestimmung des Phänomenalismus," *Vierteljahrsschrift für wissenschaftliche Philosophie und Soziologie* 36 (N.F. 11): 1–18.

—— (1913), *Der Phänomenalismus. Eine naturwissenschaftliche Weltanschauung*, Leipzig: Johann Ambrosius Barth.

Köhnke, K. C. (1997), "Neukantianismus zwischen Positivismus und Idealismus?" in *Kultur und Kulturwissenschaften um 1900. II: Idealismus und Positivismus*, edited by G. Hübinger, R. vom Bruch, F. W. Graf, Stuttgart: Franz Steiner, 41–52.

Külpe, O. (1902), "Ueber die Objectivirung und Subjectivirung von Sinneseindrücken," *Philosophische Studien* 19: 508–556.

—— (1904), "Versuche über Abstraktion," *Bericht über den ersten Kongreß für experimentelle Psychologie*, 56–68.

—— (1912), *Die Realisierung. Ein Beitrag zur Grundlegung der Realwissenschaften*, vol. 1, Leipzig: S. Hirzel.

Kusch, M. (1995), *Psychologism: A Case Study in the Sociology of Philosophical Knowledge*. London and New York: Routledge.

—— (1999), *Psychological Knowledge: A Social History and Philosophy*, London and New York: Routledge.

—— (2017), "Epistemic Relativism, Scepticism, Pluralism," *Synthese* 194: 4687–4703.

Landmann-Kalischer, E. (1905), "Über den Erkenntniswert ästhetischer Urteile. Ein Vergleich zwischen Sinnes- und Werturteilen," *Archiv für die gesamte Psychologie* 5: 263–328.

Meinong, A. (1888), "Über Sinnesermüdung im Bereiche des Weber'schen Gesetzes," in Meinong, *Abhandlungen zur Psychologie: Der Gesammelten Abhandlungen Erster Band*, Leipzig: Barth, 1914, 77–108.

—— (1896), *Über die Bedeutung des Weber'schen Gesetzes. Beiträge zur Psychologie des Vergleichens und Messens*. Hamburg and Leipzig: Voss.

—— (1911), "Für die Psychologie und gegen den Psychologismus in der allgemeinen Werttheorie," *Atti del IV congresso internazionale di filosofia* 3, 132–147.

Messer, A. (1926), *Deutsche Wertphilosophie der Gegenwart*, Leipzig: Reinicke 1926.

Münsterberg, H. (1909a), *The Eternal Values*, Boston and New York: Houghton Mifflin.

—— (1909b), "The Opponents of Eternal Values," *The Psychological Bulletin* 6: 329–338.

Perry, R. B. (1907), "A Review of Pragmatism as a Philosophical Generalization," *The Journal of Philosophy, Psychology and Scientific Methods* 4: 421–428.

Reicher, M. E. (2016), "Ästhetische Werte als dispositionale Eigenschaften: 1905–2014," in *Geschichte – Gesellschaft – Geltung. XXIII. Deutscher Kongress für Philosophie, Kolloquienbeiträge*, edited by M. Quante, Hamburg: Meiner, 961–974.

Rickert, H. (1913), "Vom System der Werte," *Logos* 4: 295–327.

Schneider, H. (1917), "The Theory of Values," *The Journal of Philosophy, Psychology and Scientific Methods* 14: 141–154.

Urban, W.M. (1907), "Recent Tendencies in the Psychological Theory of Values," *The Psychological Bulletin* 4: 65–72.

—— (1909), *Valuation. Its Nature and Laws Being an Introduction to the General Theory of Laws*, London and New York: Swan Sonnenschein, Macmillan.

Wundt, W. (1893), *Grundzüge der physiologischen Psychologie*, 2nd ed., vol. 1, Leipzig: Wilhelm Engelmann.

Ziche, P. (2008), *Wissenschaftslandschaften um 1900. Philosophie, die Wissenschaften und der "nicht-reduktive Szientismus."* Zürich: Chronos.

—— (2015), "'Gefühlsgewissheit' und 'logischer Takt.' Neue Erfahrungsmodalitäten und offene Wissenschaftsbegründung um 1900," *Scientia Poetica* 19, 322–341.

—— (2016), "Theories of Order in Carnap's *Aufbau*," *Influences on the Aufbau*, edited by C. Damböck, Berlin: Springer, 77–97.

Ziche, P. (ed.) (1999), *Introspektion. Texte zur Selbstwahrnehmung des Ichs*, Wien and New York: Springer.

Ziehen, T. (1913), *Erkenntnistheorie auf psychophysiologischer und physikalischer Grundlage*, Jena: Gustav Fischer.

—— (1924), *Leitfaden der physiologischen Psychologie in 16 Vorlesungen*, 12th ed., Jena: Gustav Fischer.

—— (1927), *Das Problem der Gesetze. Rede gehalten bei dem Antritte des Rektorats der Vereinigten Friedrichs-Universität Halle-Wittenberg am 12. Juli 1927*, Halle (Saale): Max Niemeyer.

Part III

Epistemology

Introduction

Niels Wildschut

From the moment Immanuel Kant (1724–1804) declared that "nature" was nothing but the whole of experience, and that any enquiry into nature first required a critical assessment of the necessary preconditions and limitations of experience, many philosophers began to see it as their central task to provide the sciences with their epistemological foundations. In the nineteenth century, however, the Kantian self-understanding of the philosopher was also challenged from various sides. For example, the institutionalization of history as a scientific discipline raised the question: How can the philosopher's methodologies of critique and of transcendental deduction be reconciled with the insight that all of life on earth, including the philosopher's own standpoint, is thoroughly historical? Furthermore, various natural sciences posed their own challenges to philosophy, for example, by uncovering the psycho-physiological processes within human cognition, and by stressing the very earthly and material nature of the soul (and thus presumably also of the "transcendental subject"). From within Kant's transcendental philosophy, many of these challenges seemed easy to block. Within the scientific community as well as the general (and rapidly expanding) reading public, however, these Kantian defenses appeared less and less convincing as the century progressed.

Key expressions of this tension in the nineteenth century, between philosophical epistemology and scientific practice, were the "materialism debate" and the "psychologism debate." While in the 1840s, philosophers increasingly gave up their Hegelian certainties about history as the self-realization of absolute spirit, physiologists like Karl Vogt (1817–1895) and Ludwig Büchner (1824–1899) argued that the functions of *Geist* depended on brain functions, which were themselves fully determined by laws of nature. Furthermore, Ludwig Feuerbach (1804–1872) influentially analyzed the "anthropological essence" of religion, viz., God as a reflection of the human essence. Both lines of thought were intertwined with political criticism and sparked heated controversy. From the 1860s onward, moreover, the results of physiology and experimental psychology led various epistemologists and logicians Christoph von Sigwart (1830–1904) and Ernst Mach (1838–1916) to think of the norms of reasoning as (akin to) psychological laws.

Philosophers such as Otto Liebmann (1840–1912) and Friedrich Lange (1828–1875) attempted to counter these trends with the slogan "Back to Kant!" The Neo-Kantian movement aimed to apply Kant's "scientific philosophy" to the new historical circumstances. For Hermann von Helmholtz (1821–1894), "scientific philosophy" meant that philosophical investigation needed to incorporate the results of sense-physiological experiment. Lange, furthermore, interpreted the results of physiology and experimental psychology as confirming fundamental features of Kant's transcendental psychology, including the very distinction—between phenomena and noumena—which precludes the doctrine of materialism (at least) as an ontology. Hermann Lotze (1817–1881) and Hermann Cohen (1842–1918), by contrast, defended anti-psychologistic interpretations of Kant's critical philosophy, and they sharply distinguished between (physiological or psychological) genesis and (philosophical) validity. Lotze and Cohen argued that the sciences were themselves epistemologically naïve, as scientists did not pay due respect to the manner in which experience, hence also scientific experiment, was preconditioned (and necessarily limited) by the pure forms of intuition and the categories of the understanding. In general, the Neo-Kantians claimed that since the new sciences ignored the distinction between objects of experience and things-in-themselves, they erroneously presented their speculations concerning soul, God, and free will as having the status of established scientific theories.

While the natural scientists were not overly impressed with this epistemological rejoinder from the philosophy faculty, philosophy was also under attack from other sides. Influential groups were the recently institutionalized professions of the historians, the biologists, and the social scientists. (See the Science section of this volume.) But also from the margins of philosophy itself, loud voices were being raised. The late fame of the works of Arthur Schopenhauer (1788–1860) presented the Neo-Kantians with a set of challenges: Schopenhauer's ethical conception of philosophy as chiefly dealing with the "problem of existence" gave the discipline a wider relevance which (at least for the general public) the Neo-Kantians' epistemological orientation could not provide. And Schopenhauer's position that in interpreting the meaning of existence, philosophy must affirm that it proceeds metaphysically, became the main alternative to the Neo-Kantian understanding of philosophy. Furthermore, Friedrich Nietzsche (1844–1900) confronted philosophy with a radical critique that challenged its traditional self-understanding of being committed to an impartial pursuit of truth. Nietzsche aimed to identify the philosophers' truths as nothing more than their favorite prejudices. His method of "genealogy" attempted to uncover the psychological and historical origins of philosophical convictions. Nietzsche dramatically raised many of the problems that would haunt the intellectual climate of the following decades, such as contingency, ideology, and power, as well as their philosophical correlates, historicism, nihilism, and relativism.

As the first chapter in this section stresses, Nietzsche's critical reorientation of philosophy remains a challenge, especially in interpreting his own oeuvre as well as the developments he went through. For instance, it is still heavily debated whether Nietzsche's idea of "perspectivism" has relativistic implications. According to Brian Leiter's assessment of Nietzsche's perspectivism, however, the necessarily perspectival character of human knowledge is nothing like the Protagorean "man is the measure" doctrine that it is sometimes associated with. For one, Nietzsche thinks that evolutionary pressures shape human beings' dispositions toward certain kinds of beliefs, so that their knowledge is relative only to the perspective of the human species. For another, in denying Kant's thing-in-itself any relevance to human cognition and asserting that all knowing depends on affects, Nietzsche's perspectivism states that the more affects we engage, the more we know. In Nietzsche's own inquiries into human beings and their motivations, this aspect of his perspectivism appears crucial: affectively engaging with different (types of) human beings helps to learn why they valuate and act as they in fact do.

The epistemological, existential, and political challenges posed by Schopenhauer and Nietzsche did not solely affect the new radical movements in philosophy and art that arose around 1900, such as symbolism, fauvism, expressionism, and *Lebensphilosophie*. Also, the two schools of Neo-Kantianism, and later the movement of phenomenology, became aware that the "problem of relativism" was not so easily resolved. The Marburg school's main concern was with spelling out the conditions of experience (which Cohen and others found in the concepts and methods of the mathematical sciences, and Ernst Cassirer (1874–1945) in the symbolic forms that condition the development of knowledge and culture). The Baden school, with main representatives, e.g., Wilhelm Windelband (1848–1915) and Heinrich Rickert (1863–1936), by contrast, reformulated the critical project as a "philosophy of values," which would not only determine the basis of normative judgment, but also allow to demarcate an independent realm for the historical and cultural sciences. Windelband, for instance, had discussed relativism as a philosophical problem already in 1880. He was one of the first to conceive of "psychologism" and "historicism" as inevitably leading to relativism, because of their crossing the boundary between genesis and validity. Windelband argued that all attempts by psychologicists and historicists to empirically vindicate the norms for epistemology, ethics, and aesthetics were doomed to fail, as they attempted to derive normativity from contingent phenomena.

The approach of Ernst Cassirer (1874–1945) to the problem of reconciling Neo-Kantian philosophy with the results of the historicist tradition is discussed by Samantha Matherne in the second chapter. Matherne argues that although Cassirer rejects "complete relativism," he endorses another form of relativism that is grounded on Kantian principles, which Matherne labels "critical relativism." According to Cassirer, the complete relativist defines objectivity as correspondence to something absolute while simultaneously denying that absolutely true judgments are possible. As a consequence, no objectively true judgments are possible, and truth is relativized to individual subjects. Critical relativism, on the other hand, defines objectivity in terms of conforming to the functions (concepts, laws, principles) of experience. Thus, while also denying that judgments could be absolutely true, the critical relativist still thinks that they can be measured according to the ideal limit of the "whole" of cognition. Central to Cassirer's argument in favor of critical relativism, then, is the claim that our judgments in morality are like those in natural science: their objectivity may be measured according to the extent in which they govern the wholes of experience and of willing respectively. In this way, the idea of progressing science and morality plays a crucial role in Cassirer's understanding of the relativity of both.

In the context of grounding sociology as a science and, specifically, establishing the "sociology of knowledge," historicism and relativism presented not merely profound challenges, but also great opportunities. As Martin Kusch highlights in the third chapter, Georg Simmel (1858–1918) and Karl Mannheim (1893–1947) both adopted relativism, at least as a method for uncovering the social, material, and historical conditions of (philosophical) knowledge. Strikingly, in their sociological-historical case studies, they traced the emergence of "relativism" and "historicism" as products, e.g., of modern monetary reality (Simmel) and of the tradition of political conservatism (Mannheim). They thus turned the historical reflexivity, mentioned at the beginning of this introduction, from a threat to epistemology into a methodological virtue of their sociological research programs. In addition, Kusch analyzes how Simmel and Mannheim dealt with the philosophical implications of their sociology. Simmel declared himself a relativist and aimed to relativize even the opposition between relativism and absolutism. However, assessing whether Simmel actually was a relativist, either by the standards of the time or by our standards today, remains a challenge. Mannheim, by contrast, thought the sociology of knowledge had to avoid relativism at all costs—but Kusch concludes that his attempts to do so failed.

The Neo-Kantian and phenomenologist relativism-charge did not merely affect those who aimed to follow the sciences. The charge also returned to haunt those who attempted to radically revise the rules of philosophy and metaphysics. Thus, Edmund Husserl (1859–1939) also called out Martin Heidegger (1889–1976) for solely focusing on the "essence of human being's concrete worldly Dasein" and for thus committing to "anthropologism," psychologism, and indeed, (species-)relativism. In Sacha Golob's chapter on the question whether Heidegger is in fact a relativist, Golob assesses Husserl's charge by countering a sophisticated interpretation of Heidegger as a relativist. According to this relativistic reading, Heidegger is a "conceptual scheme" relativist who historicizes the *a priori* and claims that multiple incommensurable frameworks exist for understanding being. By contrast, Golob argues that Heidegger offers a hermeneutic analysis of the underlying assumptions of the opposition between relativism and absolutism. Importantly, Golob interprets Heidegger as a realist, for whom the essential task of phenomenology and hermeneutics consists in adjusting and recalibrating our standpoint, so that we "arrive at the right perspective."

8 Knowledge and affect

Perspectivism reconsidered

Brian Leiter

Perspectivism: The interpretive dilemma

Nietzsche's idea of "perspectivism" has often been equated with relativism, with the idea that there is no objective knowledge, only "knowledge" relative to a perspective; and that there is no objective truth, only "truth" relative to a perspective.[1] Call this "Perspectivism as Protagoreanism" (hereafter PaP), since the relativism at issue echoes the famed Protagorean "man is the measure" doctrine: What is true for me, may be false for you, and vice versa. The PaP reading, as we will see, finds little support in what Nietzsche actually writes about perspectivism, and it also generates a serious interpretive dilemma (Leiter 1994, 336–338): How, on PaP, would we make sense of Nietzsche's many claims that appear to presuppose an *epistemic privilege* on his part, i.e., that he knows something (not relatively, but objectively) that the targets of his criticisms do not?

For example, Nietzsche frequently claims that genuine knowledge comes from the senses (but not from other sources); that Christian and Platonic metaphysics are actually false; that naturalistic explanations are epistemically superior to moral and religious ones; and that certain purported truths are really errors. As I argued more than two decades ago (Leiter 1994), Nietzsche makes consistent use of what I called "epistemic value" terms (e.g., truth/lie; true/false; *wirklich/unwirklich* [real/unreal]) in defending his views and criticizing others: this choice of language makes no literal sense without Nietzsche assuming that he enjoys privileged epistemic access to non-relative truths. Many relevant passages are reviewed in Leiter (1994, 336–339).

In Leiter (1994), I suggested two possible ways of resolving this interpretive dilemma. The one I adopted then (following Maudemarie Clark [1990]) was to argue that, in fact, Nietzsche did not accept any kind of relativistic or skeptical doctrines about knowledge or truth (i.e., doctrines that deny the *objectivity* of knowledge or truth), except in early unpublished work and for reasons he came to repudiate. Somewhat later, on Clark's reading, Nietzsche expressed a different kind of anti-objectivist view about knowledge and truth based on his acceptance of the phenomenal/noumenal distinction, and his assumption, like many early critics of Kant, that by limiting knowledge to the merely phenomenal world, Kant had denied humans knowledge of objective reality. But Nietzsche later abandoned that distinction on the grounds that the idea of a "noumenal" world was unintelligible. If that were right, then knowledge of the "phenomenal" world was the only knowledge to be had, thus explaining the mature Nietzsche's confidence in knowledge delivered by the empirical sciences.

In more recent work, I have argued (Leiter 2002, 2015a) for a "pragmatic" reading as a partial alternative to Clark's developmental hypothesis.[2] Even if, *contra* Clark, it *were* intelligible that there is a way things are independent of all perspective (i.e., a noumenal world), such a reality would not be of any practical concern or relevance to us: a reality beyond our ken (our cognitive capacities) is *of no practical interest or significance*. (The famous passage from *Twilight of the Idols* on *Wie die 'wahre Welt' endlich zur Fabel wurde* [How the "True World" Finally Became a Fable], understood as a description of the evolution of Nietzsche's own views, suggests precisely this at its conclusion, as John Wilcox first argued nearly a half-century ago [Wilcox, 1974]: for discussion cf. Leiter [2015a, 221–223].)

In my earlier essay on perspectivism (1994), I did acknowledge another possible response to the interpretive dilemma, one that still awaits a serious defender (assuming the position can be defended):

> One might grant that Nietzsche makes criticisms with epistemically loaded language (truth, lie, reality, false, and so on), but dispute that it should be interpreted literally. It might be argued that he uses this language for rhetorical effect only (it helps persuade to have "truth" and "reality" on one's side). Such a reading might emphasize the possible Sophistic pedigree of Nietzsche's epistemological views, and claim that, in best Sophistic fashion, he appreciates the rhetorical value of epistemically loaded—but semantically empty—language.
>
> (Leiter 1994, 339)

The possibility that Nietzsche's extensive use of epistemic value terms in defending his views and criticizing those of others constitutes "mere rhetoric" is compatible with PaP. We know that Nietzsche was an admirer of Protagoras, though often using him as simply a named placeholder for anti-Platonic views quite generally (cf. Mann and Lustila 2011). Protagoras, and his "man is the measure" doctrine, is treated very favorably as well by Friedrich Lange in his *History of Materialism*, a book that had a huge impact on Nietzsche after he discovered it in 1866 (though Lange's Protagoras is an empiricist and proto-Kantian, whose primary difference from Kant is that he denies that the subject's contribution to experience is universally shared, a point that will prove important in understanding Nietzsche's view, as we shall see).[3]

We should remember, however, that Nietzsche's primary symbol of Sophistic culture was always Thucydides, not Protagoras—the Thucydides who, as Nietzsche puts it, takes "delight in all that is typical in men and events and believes that to each type there pertains a quantum of *good sense*: *this* he seeks to discover ... Unlike Plato he does not revile or belittle those he does not like or who have harmed him in life. On the contrary: through seeing nothing but types he introduces something great into all things and persons he treats of ..." (D 169). Some commentators suggest that Thucydides "revels in the richness of these human type perspectives" (Mann and Lustila 2011, 56), and thus evokes Protagoras's "man is the measure" doctrine (except it is the *type* of man who is the measure, not the individual perceiver). There is something importantly right about this, though it appears, on its face, initially hard to square with Nietzsche's praise of Thucydides as symbolizing that "*culture of the most impartial* [or 'unprejudiced': '*unbefangensten*'] *knowledge of the world*" one "which deserves to be baptized with the name of its teachers, the Sophists" (D 168). I think there is a way of resolving this apparent tension, which will help illuminate Nietzsche's perspectivism, and show it to be more influenced by Thucydides than Protagoras. We will return to Thucydides at the end, but let us start by looking carefully at the two actual texts in the published corpus where Nietzsche addresses "perspectivism" at length.[4]

Nietzsche's published texts on perspectivism

Perspectivism is discussed in detail in only two passages in the works Nietzsche chose to publish: section 354 of *The Gay Science* (GS 354) and the Third Treatise, section 12 of *On the Genealogy of Morality* (GM III:12). The term itself, *Perspektivismus*, is used only in GS 354. GM III:12 uses several cognate terms—*Perspektiven* (perspectives), *perspektivisches Sehen* (perspective seeing), *perspektivisches Erkennen* (perspective "knowing")—but never *Perspektivismus*. GS 354 was part of the Fifth Book added to the second edition of *The Gay Science* some five years after the first four "books" of GS appeared (1881); in other words, the Fifth Book was written around the same time as GM III:12 (1886). One might hope an interpretation of perspectivism could show how these two passages fit together, and I will argue that they do.

Briefly, GS 354 claims that our conscious "knowledge" of the world is influenced by evolutionary pressures: what we take ourselves to know is only that which serves the needs of the species, as it were. Such knowledge is relative to perspective, but this is the *human perspective,* one we cannot escape in virtue of being human. I will call this *Global Humeanism,* since it supposes, with Hume, that the non-rational dispositions toward forming certain kinds of beliefs are so widely shared (due to evolutionary pressures on Nietzsche's view, not Hume's, obviously) that they are *typical of being human* and thus give rise to a common body of beliefs about the world that are socially useful, even if, strictly speaking, false (or unwarranted).[5] By contrast, GM III:12 advances a superficially different claim, namely, that all knowing is dependent (in some sense) on affects. GM III:12 advances this thesis, in particular, against the possibility Schopenhauer held out of affect-free knowledge: for Nietzsche, all knowing—even of space, time and causality—is dependent on the affects.

This will form the crux of the continuity between the passages, as I will argue below. In the end, I will argue, GM III:12 only demonstrates the role that affects play in what comes to our cognitive attention, not in constituting what really exists. Now let us turn to the central texts in more detail and a more detailed exposition of the preceding summary.

The gay science

GS 354 begins with a puzzle: Why should thinking, feeling, willing, and remembering be conscious, since each could transpire without being conscious, without "seeing itself in a mirror" (as Nietzsche puts it)?[6] The answer Nietzsche proffers is that these mental states become conscious because of the need to communicate with others: "consciousness is ... a net of communication between human beings; it is only as such that it had to develop." Here, Nietzsche echoes Johann Gottfried Herder's view that conscious thinking is articulated in language (as Nietzsche puts it: "the development of language and the development of consciousness ... go hand in hand"), but (this is Nietzsche, not necessarily Herder) language develops only as "required by social or herd utility," which consists in communicating what is common in experience (not what is distinctively individual). Nietzsche equates this with "perspectivism": "all becoming conscious involves a great and thorough corruption, falsification, reduction to superficialities," i.e., the superficialities conducive to socially useful communication (superficial precisely because they pick out only what is common to different experiences and thus useful for social coordination: cf. BGE 268 for a similar argument). In context, it seems clear that Nietzsche does not mean this point to apply to brute perceptual consciousness (e.g., experiencing redness or sweetness) but rather to the linguistically articulable awareness of our own mental life, or the mental lives of others, as well as observable features of the world we inhabit.[7] His conclusion: "We simply lack any organ for knowledge, for 'truth': we 'know' (or believe or imagine) just as much as may be *useful* in the interests of the human herd, the species ..."

There are two key ideas at the conclusion of GS 354. First, Nietzsche recognizes that evolutionary forces affecting human cognition do not necessarily prefer true to false belief—even in the case of ordinary knowledge about the empirical world (cf. Stich 1978, 55–60, for a seminal discussion: natural selection favors false positives or false negatives over accuracy in perceptual reports whenever the former confer reproductive advantage). This point will be especially important with regard to claims about our mental lives, since these are only useful for social coordination to the extent that they identify *common* features, not individual or idiosyncratic ones: what my pain upon being bitten by the snake is *really* like is irrelevant compared to the fact that *you want to avoid it!*, which is what others in my community need to "know." Nietzsche's claim is that our conscious knowledge is subject to evolutionary pressures which are only accidentally truth-tracking but are essentially reproductive-fitness-tracking: thus, we lack a capacity for know-ledge (without quotation marks), having only an ability to "know" what is useful for the "herd" (since Nietzsche takes the average type of person to be the upshot of natural selection). Call this aspect of Nietzsche's view in GS 354 "Skeptical Darwinism." As he puts it elsewhere, errors that "proved to be useful and helped preserve the species" persisted, since "those who hit upon or inherited these had better luck in their struggle for themselves and their progeny" (GS 110). Examples of such (alleged) errors include "that there are enduring things; that there are equal things; that there are things, substances, bodies; that a thing is what it appears to be; that our will is free; that what is good for me is also good in itself" (GS 110).[8] From a certain kind of rad-ically naturalistic view, common in the nineteenth century and familiar to Nietzsche, most of these claims were, indeed, false or unwarranted.

Second, Nietzsche is a *Global Humean*: he assumes that these evolutionary pressures operate to produce massive convergence on certain "errors," the ones essential to human coping with the world and with our fellow humans.[9] Thus, in the preface of *Beyond Good and Evil*, Nietzsche calls "perspective, the basic condition of all life."[10] In the final book of *The Gay Science,* added in 1886, he goes further, noting that whether all existence is perspectival "cannot be decided even by the most industrious and scrupulously conscientious ana-lysis and self-examination of the intellect; for in the course of this analysis the human intellect cannot avoid seeing itself in its own perspectives, and *only* in these. We cannot look around our own corner" (GS 374). In consequence, "we cannot reject the possibility that the world *may include infinite interpret-ations*" but that is neither here nor there for essentially the pragmatic reasons articulated in the famous *Twilight* passage on the fable of the "true" world noted above: interpretations that are unavailable to us are practically irrele-vant. There is an obvious resonance here with the twentieth-century idea of "Neurath's boat," popularized in the anglophone literature by Quine: we can only ask and answer questions about what there is and what we know from within some existing interpretation of the world, which means there is no neu-tral standpoint from which we can rule out radically different interpretations, let alone adjudicate among them. There may be radically different interpret-ations, of course, but that can hardly matter for those of us who cannot pos-sibly occupy the perspective from which they arise. And what perspectives we can, in fact, occupy are determined by the evolutionary limits set on cognition.

In sum, Nietzsche's Global Humeanism, with its Skeptical Darwinian com-ponent, must set the baseline against which all our other epistemic pursuits proceed. GM III:12, as I will argue, is compatible with that hypothesis.

On the genealogy of morality

GM III:12 appears, initially, to be a different passage about perspectivism, but it is important to recall its context within the argument of the Third Treatise of the *Genealogy*. The main topic of the Third Treatise is *why* ascetic ideals—ideals of self-denial ("chastity, poverty, humility"), ideals that "deny life"—have been so appealing to human beings, figuring, as they do, in all the world's major religions. Against those who think this asceticism is a mere legacy of religious superstition, now abolished by a scientific age, Nietzsche claims—in what is plainly intended as a dramatic surprise of the essay—that *even* science, with its faith in the absolute value of truth, also embodies the ascetic ideal (cf. GS 344 for a similar claim, also from Book V; and see Leiter [2015a, 213 ff.] for detailed discussion: Nietzsche's idea, in brief, is that error is a necessary condition of life, so the overvaluation of truth is a threat to life itself).

Well before Nietzsche defends that latter hypothesis in the Third Treatise (in GM III:23), however, he also claims that Kant's transcendental idealism expresses the ascetic ideal, because of its doctrine that "*there is* a realm of truth and being, but reason is firmly *excluded* from it!" (GM III:12). Kant's doctrine is ascetic, on this account, *not* because it overvalues truth but because it denies that we have any access to the real or objective truth about the world as it is in itself: this is supposedly ascetic because it treats as mere appearance the only living world there is (the "phenomenal" world in Kant's terms). So, whatever Nietzsche's perspectivism amounts to, it cannot mean that "the world" is cognitively off limits, lest it fall prey to the asceticism of Kant's view. This latter point is crucial to any reading of GM III:12.

Nietzsche's discussion of perspectives (GM III:12) is then introduced as an alternative, non-Kantian and non-ascetic way of thinking about "objectivity," one in which objective knowledge is *not* a matter of knowledge of a noumenal world.[11] Nietzsche writes:

> To see differently in this way for once, the *will* to see things differently, is no small discipline and preparation of the intellect for its coming "objectivity"—understood not as "contemplation without interest" (which is, as such, a non-concept and an absurdity), but as *having in our power* our "pros" and "cons": so as to be able to engage and disengage them so that we can use the *difference* in perspectives and affective interpretations for knowledge. From now on, my philosophical colleagues, let us be more wary of the dangerous old conceptual fairy-tale which has set up a "pure, will-less, painless, timeless, subject of knowledge" [quoting Schopenhauer] ... here we are asked to think of an eye which cannot be thought at all, an eye turned in no direction at all, an eye where the active and interpretive powers are to be suppressed, absent, but through which seeing still becomes a seeing-something, so it is an absurdity and non-concept of the eye that is demanded. There is *only* a perspective seeing, *only* a perspective "knowing"; the *more* affects we allow to speak about a thing, the *more* eyes, various eyes, we are able to use for the same thing, the more complete will be our "concept" of the thing, our "objectivity." But to eliminate the will completely and turn off all the affects without exception, assuming we could: well? Would that not mean to *castrate* the intellect?

To begin, notice that this passage claims *only* that *knowing* is perspectival, not that truth is. That leaves open the possibility that there is a non-perspectival truth about the world, though the pragmatic reading mentioned earlier would explain why Nietzsche thinks such a truth would be irrelevant *if it transcended all possible* perspectives creatures like us could adopt. But Nietzsche's point in GM III:12, I suggest, is far more banal and entirely consistent with his dismissal of the "noumenal world" as a symptom of asceticism: namely, that the world is sufficiently complex that the parts we attend to for purposes of acquiring knowledge are determined by our affects or interests, thus the more of them that are brought to bear the more we would know about the world (the more truths about the world we will know). That certainly fits with the visual analogy that Nietzsche offers: if I view an object from many different angles and perspectives, I will, indeed, acquire more first-hand visual knowledge of it. (The visual analogy would also imply that some perspectives do not give us knowledge [a point I emphasized in Leiter, 1994], but Nietzsche does not say that explicitly—indeed, all he says explicitly is that the more perspectives that are deployed, the greater our "objectivity" [Nietzsche's quotes]). I think this banal point is, in fact, central to Nietzsche's meaning: the world is overflowing with possible cognitive targets, and affects determine which ones we pick up upon on any given occasion. Let us call this Nietzsche's "Busy World Hypothesis": there is so much we *could* know, that what we do end up cognizing depends on our affects and interests. The Busy World Hypothesis will both help us understand why Thucydides is Nietzsche's paradigm of a perspectival knower, as it were, and how GM III:12 is, in fact, continuous with GS 354. I return to both points shortly, but first let us press further on the analogy that is central to GM III:12.

Knowing is like seeing, according to Nietzsche, in that knowing, like seeing, is *dependent* (in some sense to be specified) on a perspective (an *interest* or *affect*). Christopher Janaway has argued that affects are "constitutively necessary conditions of the knower's knowing anything at all" (Janaway 2007, 212), but this talk of dependence (or "constitutively necessary conditions") is ambiguous between two possibilities, which are apparent in the case of visual perspectives. I cannot have direct knowledge (by "acquaintance," as it were) of what the bottom of the table looks like unless I lie on the floor and look up at it: my perspective "constitutes" what I know in the sense that I could not be acquainted with how the bottom of the table appears except by occupying the relevant perspective.[12] Call the latter a case of "epistemic constitution": the perspective makes a particular bit of cognition of an objective feature of the world possible.

But there is a stronger constitutive sense of a visual perspective, in which it does not simply make "knowledge" (or awareness of [acquaintance with] some aspect of the world) *possible*, but in which its object depends *for its existence* on the perspective from which it is seen. The mirage of an oasis in the distance that appears to the man lost in the desert and dying of thirst is constitutively dependent on the perceiver in this second, "metaphysical" sense: only for that perceiver, under those circumstances and in those conditions, does the object exist. When perspectives *metaphysically constitute* an object, there is, in reality, no such object, and so any "knowledge" or awareness is illusory. Values are, for Nietzsche, metaphysically constituted in this way, since Nietzsche is an anti-realist, someone who denies that there are any mind-independent (or judgment-independent) facts about value. (For detailed discussion of the textual evidence, see Leiter 2000; 2013; 2014; 2015a; 118–125.) And this explains, why, as I have argued previously, Nietzsche never uses what I have called "epistemic value" terms (i.e., terms that designate the epistemic standing of particular judgments, e.g., true/false, warranted/unwarranted, justified/unjustified) with respect to judgments about questions of value. In other words, values are like hallucinated "Oasis Facts" in a crucial respect: their existence is relative to the person, or the type of person, who perceives them. There may be an objective (psychological) fact that events have particular value for particular persons, but there is no objective or observer-independent fact about the *real* value of events.

Of course, if *no values* are mind-independent, then that must be true of *epistemic values* as well: judgments about what is warranted or justified to believe must also be mind-dependent. I see no reason to think that Nietzsche would deny that: indeed, it is the whole point of his Global Humeanism that "creatures like us" have a tendency to deem justified certain kinds of claims about the world, even if they are not justified from a non-relative point of view. And Global Humeanism, recall, is one of the upshots of GS 354, with herd-reproducing forces explaining the epistemic values and the resulting beliefs that triumph. But since we cannot do without most of these epistemic values—since we cannot occupy any other perspective—they form the baseline for all our epistemic pursuits. This fits with a point that Christopher Janaway has correctly emphasized, namely, that GM III:12 is clearly referring to Schopenhauer. In particular, while Schopenhauer agreed with Kant that the knowing subject imposes upon experience various forms and categories like space, time, and causality, Schopenhauer, unlike Kant, deemed even these to be products of the human "will," that is, of "human needs, interests, and affects" (Janaway 2007, 193). That view is suggestive, obviously, of the doctrine I have been calling "Global Humeanism," the one we find in GS 354 as well. As Janaway goes on to note,

> Schopenhauer polarizes will and intellect, and likes to tell us that while the brain is the focus of the intellect, the genitals are the focus of the will. Hence when Nietzsche speaks of "castrating the intellect" [in GM III:12] he makes a direct assault on Schopenhauer's aspiration towards a will-free operation of the intellect: we must accept the intellect as essentially will-driven …
>
> (Janaway 2007, 200)

In other words, when GM III:12 rejects the possibility of Schopenhauer's "pure, will-less, painless, timeless, subject of knowledge" he is affirming that even the Schopenhaurian intellect is "essentially will-driven" (i.e., affect-driven): there is *contra* Schopenhauer, never any "contemplation without interest." Thus, Nietzsche leaves us with the view that the only knowing possible for us—including that involved in the Kantian intuitions of space and time and the categories of the understanding—is epistemically dependent on human interests and affects.

And now the continuity between GS 354 and GM II:12 should be clear: both passages presuppose the Busy World Hypothesis, and thus that what we know "depends" (in the epistemic sense) on affects—both our knowledge of the familiar Kantian phenomena (e.g., space, time, causality), but also our knowledge more generally. Both GS 354 and GM III:12 involve a kind of naturalized Kantianism, or what I have been calling Global Humeanism: whereas Kant treats the uniformity of human judgment on these matters as transcendental conditions on the possibility of knowledge, Nietzsche treats them as simply consequences of our cognitive propensities as shaped by evolutionary forces. However, GM III:12 also calls attention to the role affects play in knowledge beyond the Kantian case, i.e., the cases where we can turn on, and turn off, as it were, *certain* affects in the course of inquiry. And that is crucial for understanding the import of perspectivism in GM III:12. But before getting there, I want to consider an alternative interpretive hypothesis.

Janaway's circumscribed perspectivism

"Willing" for Schopenhauer, includes "all desiring, striving, wishing, longing, yearning, hoping, loving, rejoicing, exulting, and the like, as well as the feeling of unwillingness or repugnance, detesting, fleeing, fearing, being angry, hating, mourning, suffering, in short, all affects and passions [*Affekte und Leidenschaften*]" (from the *Prize Essay on the Freedom of the Will*, quoted in Janaway 2007, 202). Janaway takes affects to be "at bottom inclinations [*Neigungen*] or aversions [*Abneigungen*] of some kind" (Janaway 2007, 205; cf. Janaway 2007, 205–206 for a catalogue of the kinds of affects Nietzsche mentions), and following Janaway, I have argued (Leiter 2013) that the *basic* affects of inclination and aversion are for Nietzsche non-cognitive:[13] they are individuated by their phenomenological feel, and while they have intentional objects ("feeling towards" [Goldie 2002]), they have no truth-evaluable content. I will not rehearse the arguments for that here (see Leiter 2013).

Janaway proposes that GM III:12 is not "a generalization about all knowing"—as the Global Humeanism of *The Gay Science* passage appears to be—but applies only to cases like "knowledge about the various phenomena of morality" (Janaway 2007, 209), the obvious subject of the *Genealogy*. Of course, when Nietzsche repudiates the possibility of "contemplation without interest," he is also repudiating a possibility related to purported aesthetic knowledge or appreciation (Kant's central case), suggesting that the target is broader than just morality and encompasses other claims about value. Yet, there is nothing in GM III:12 itself that circumscribes its claim about cognition to evaluative phenomena—and, as I will argue below, Nietzsche's paradigmatic knower, Thucydides, illuminates primarily psychological phenomena, not ethical or aesthetic ones. And if affects are non-cognitive, and they even play a role in, e.g., knowledge about causation, then it is hard to see why GM III:12's claims should be circumscribed in the way Janaway proposes.

Janaway argues that affects are "constitutively necessary conditions of the knower's knowing anything at all" because "even if scientific investigation has to be construed as a form of knowledge purged of all affects" it could still be that "knowing something *only* scientifically gives us a poorer understanding of it than knowing it through a variety of psychological, imaginative, rhetorical means—affect-arousing means—in addition to those of science" (Janaway 2007, 212). Janaway's idea seems to be that we cannot really understand the psychological truths about morality (e.g., that it arises from *ressentiment*, that it manifests internalized cruelty) that Nietzsche's genealogy reveals unless we are affectively engaged by them. That is hardly obvious, for reasons I will return to in a moment. But I also want to take issue with Janaway's contrast between *Wissenschaft* and Nietzsche's own method. Any method that reliably produces knowledge of what is true is a *Wissenschaft*, and Janaway and I both agree that Nietzsche takes his method in the *Genealogy* to deliver such knowledge. The relevant contrast is not with "science" *per se*, but with science as practiced by "scholars."

Recall that Nietzsche concedes in *Beyond Good and Evil* (BGE 6) that *Gelehrte* (scholars), those who are really scientific (*eigentlich wissenschaftlichen*) men may have an actual *Erkenntnistrieb* (knowledge drive), such that one can wind them up like a clock, and they get to work acquiring knowledge: "it is almost a matter of total indifference … whether the 'promising' young worker turns himself into a good philologist or an expert on mushrooms [*Pilzekenner*] or a chemist." This is because, according to Nietzsche, the scholar's real affective attachments lie elsewhere: "in his family, or in making money, or in politics." As Janaway puts it: Nietzsche "regards it as possible, in exceptional cases, to pursue knowledge in disconnection from the sum of one's affects and drives, but this type of 'little clockwork mechanism' cannot be the basis of a general theory of knowing, nor an accurate guide to the nature of knowing subjects" (Janaway 2007, 216). The "clockwork" case is not, however, exceptional; universities suggest it is the norm, as I am sure Nietzsche would agree (cf. GS 373). But the key point is that Nietzsche thinks there is genuine *wissenschaftlich* activity aiming at knowledge that affects do *not* even epistemically constitute—except in the explanatorily otiose sense that they reflect an *Erkenntnisstrieb*—though affects do influence it indirectly and non-constitutively, insofar as the *Erkenntnisstrieb* operates in service of "family, or in making money, or in politics." The affective arousal required for knowledge of morality is still part of a *Wissenschaft*, just one fundamentally different from that of scholars.

According to Janaway's ambitious hypothesis, affective arousal illuminates aspects of the object (i.e., morality) that a merely "scientific" understanding misses. More precisely, on Janaway's reading, we really do *not know* what morality is like unless our affects of disgust, contempt, and anger are aroused by Nietzsche's genealogy of our morality. But why should that be true? In what sense do we not "know what morality is" absent such arousal? This is the central issue for Janaway's attempt to circumscribe perspectivism.

Perhaps an analogy with the Freudian hypothesis about transference and its role in therapy might help. On Freud's view, the psychoanalyst can tell the patient that he has an unconscious wish to kill his father, and even adduce all the relevant evidence from the patient's behaviors, relationships, dreams, jokes, or slips of the tongue that supports that hypothesis. And yet, on the Freudian view, the patient does not *really* know he harbors such an unconscious wish until he transfers the unconscious murderous rage at his father on to the analyst, until he consciously experiences the wish *transferred* upon the analyst. We might say that, absent transference, the patient only acquires what Bertrand Russell used to call "knowledge by description" or "propositional knowledge," rather than "knowledge by acquaintance."

I propose a slight modification of the Russellian categories to capture what is at stake here. Let us say that an agent has propositional knowledge when he believes that X and has good reasons for affirming such a belief. Let us say that an agent *Affectively Knows* X when he has either propositional knowledge of X or becomes acquainted with X and, as a result, is motivated to act on his belief. When an agent *Affectively Knows* some proposition, he has undergone what I will call "belief fixation." Not everyone who affirms the proposition X has undergone belief fixation: the hallmark of belief fixation is that the agent who affirms X is motivated to act upon his belief X. The key to Freudian transference is belief fixation, i.e., that the patient *Affectively Knows* some proposition. This is what is at issue in the therapeutic context: it is not enough for the patient to believe (propositionally) that he has a particular unconscious wish; he must believe it in such a way that he is then *motivated* to act, to reject the actual wish, accede to it, take steps to repress it, and so on.

Might this framing of the issues help Janaway's case for limiting GM III:12 just to knowledge of morality? Janaway's claim is that when Nietzsche arouses the readers' affects toward morality, they come to know something different, namely, something about morality's origins, motivations, and real nature, things they otherwise would not have known. But all that the Freudian analogy shows is that agents would not have acquired Affective Knowledge in the absence of affective arousal, not that they would not be able to understand Nietzsche's hypotheses about the psychological origins of morality. Some readers, after all, are convinced by Nietzsche that the slaves feel *ressentiment* toward the masters based on the evidence in GM I, and conclude that the slaves are quite right to feel that way, and celebrate their victory. It seems like one can have propositional knowledge of the claims Nietzsche is making in his *Genealogy* without sharing in Nietzsche's contempt for slave morality or his desire to throw off its shackles.

I have argued for many years (e.g., Leiter 2002, 115–136, 176–180) that Nietzsche's primary aim is to transform the false consciousness of some of his readers—nascent higher human beings—about the dominant morality: he wants them to see that what passes for morality is actually *bad for them,* and thus inspire them to throw off its shackles. Nietzsche, we might say, aims at producing knowledge about morality that *motivates* action, and the only kind of knowing that does that is propositional knowledge (or acquaintance) accompanied by an affect (or, as a Humean might say, by a desire): that is, Nietzsche aims to produce Affective Knowing, at least among select readers. Mere "scholars" in Nietzsche's sense do not aim for Affective Knowing: what they know is a matter of practical indifference.

But this does not help Janaway's reading, since Affective Knowing is not what GM III:12 is actually about. Recall the crucial recommendation of the passage, namely, that *"having in our power* our 'pros' and 'cons'… so as to be able to engage and disengage them so that we can use the *difference* in perspectives and affective interpretations for knowledge."* GM III:12 recommends an approach that contributes to knowledge (no quotation marks this time), not to motivation or action. And on this reading, there are facts about morality's origins and motivations that transcend particular perspectives, but which an aggregation of affective perspectives will illuminate. In this way of thinking about the perspective metaphor, there is a way in which things are independent of any particular perspective (albeit what that fact is will be constrained by Global Humeanism), and if we maximize our epistemic angles on the moral phenomena, we will increase our "objectivity." (Nietzsche puts "objectivity" in quotes in GM III:12 precisely because he is denying, as Janaway has plausibly argued, the Schopenhauerian view that cognition is epistemically independent of the will or affect.)[14] Nietzsche's admiration of Thucydides is, I believe, the key to understanding what he is after, since Thucydides, as Nietzsche understands him, is the paradigm of knowing that GM III:12 describes. We shall see that the kind of psychological knowledge that Thucydides provides depends on affects in a far weaker sense than that of producing Affective Knowledge, and much closer to what the Busy World Hypothesis should lead us to expect.

Thucydides and perspectivism

Thucydides, Nietzsche's primary representative of Sophistic culture, did not embrace anything like PaP, nor did Nietzsche believe he did. Recall that Nietzsche praised Thucydides as symbolizing that *"culture of the most impartial* [or 'unprejudiced': *'unbefangensten'*] *knowledge of the world"* one "which deserves to be baptized with the name of its teachers, the Sophists" (D 168). Nietzsche's admiration of Thucydides in these respects is consistent with his own perspectivism in both *The Gay Science* and the *Genealogy* senses. The "knowledge of the world" Thucydides' *History* conveys is obviously *not* knowledge of the physical or biological world, but rather a special kind of psychological knowledge, namely, *about human beings and their motivations*: he illuminates what they *value*, what events *mean* to them, and thus why they act as they do. Thucydides is "impartial" in that he is willing to accurately describe their motivations and values, even when they are offensive to conventional moral sensibilities, or even to Thucydides' own judgment of the merits. Thucydides can do this precisely because he can "engage and disengage" his own "pros and cons"—his own sympathies and antipathies—in order to enhance knowledge of why the actors in the Peloponnesian War did what they did. Unlike Plato, as Nietzsche notes, Thucydides "does not revile or belittle those he does not like..." (D 169).[15] He portrays them as they are.

Thucydides' standard device in the *History* is to put into the mouths of his speakers their *real* intentions and concerns, even if they would be shocking or offensive to observers. The classic example is the famed speech that, in Thucydides' fictional recounting, the Athenians deliver to the vanquished Melians:

> For our part, we will not make a long speech no one would believe, full of fine moral arguments—that our empire is justified because we defeated the Persians, or that we are coming against you for an injustice you have done to us ... Instead, let's work out what we can do on the basis of what both sides truly accept: we both know that decisions about justice are made in human discussions only when both sides are under equal compulsion [i.e., only among equals does right prevail over might]; but when one side is stronger, it gets as much as it can, and the weak must accept that ... We did not make this law, and we were not the first to follow it; but we will take it as we found it and leave it to posterity forever, because we know that you would do the same if you had our power, and so would anyone else.[16]

Nietzsche's own commentary on this particular dialogue highlights what he admires about Thucydides' rendering of the event:

> Do you suppose perchance that these little Greek free cities, which from rage and envy would have liked to devour each other, were guided by philanthropic and righteous principles? Does one reproach Thucydides for the words he puts into the mouths of the Athenian ambassadors when they negotiated with the Melians on the question of destruction or submission?
>
> Only complete Tartuffes [i.e., Socrates and Plato] could possibly have talked of virtue in the midst of this terrible tension ...
>
> Grote's tactics in defense of the Sophists are false: he wants to raise them to the rank of men of honor and ensigns of morality—but it was precisely their honor not to indulge in any swindle with big words and virtues—.

(WP 429)

Socrates and Plato are not impartial: they let their moral indignation color their perception of what was going on around them; Thucydides, although he surely disapproves of the unbridled lust for glory and power that leads Athens to ruin (which is the story his *History* tells), nonetheless resists moralized denunciations in favor of exposing the actual motivations that drive the Athenians. The Athenians, after all, did not speak to the Melians as Thucydides portrays them, just as the victors in every war rarely speak so directly to the vanquished. All talk the language of "justice" and "fairness" and "philanthropic and righteous principles," but what Nietzsche admires about Thucydides is that he puts into the mouths of his Athenian victors *their actual beliefs and motivations*—hence Thucydides' "impartiality." In psychological reality, self-interest and lust for power drive everyone in this dialogue: the Melians think there is a question of justice at stake, since "justice" is the only card they have to play having been defeated; the Athenians may talk the talk of justice, too, but in reality they think exactly as Thucydides portrays them, given their superior military position. Thucydides honors the perspectivism dictum (of GM III:12) of "allow[ing] ... more eyes" to speak about the event in order to increase "objectivity" precisely by recording the way in which the same event appears differently to the differing types whose perspectives he records. Are the Athenians really behaving unjustly toward the Melians? Are the Melians really demanding genuine justice from the Athenians? The answer to both questions is presumably negative for Nietzsche, given his anti-realism about value. The singular talent of Thucydides is his ability to represent how the meaning of what is transpiring *really* seems to each of the opposed parties.

A different passage, later in the Third Treatise of the *Genealogy,* helps illuminate the kind of perspectivism Thucydides exemplifies. Nietzsche there says that "doing violence, pressing into orderly form, abridging, omitting, padding, fabricating, falsifying ... belong to the *essence* of all interpreting" (GM III:24). Superficial readers sometimes infer from passages like this— with its reference to "all interpreting"—that Nietzsche is committed to something like PaP about all knowledge and truth, yet the context of this remark is of decisive importance in understanding perspectivism, since its claim only purports to describe what "interpreters" actually do. Recall that the "ascetic priest" was said by Nietzsche to have interpreted "a piece of animal psychology" (GM III:20), namely, the internalization of cruelty which first gave rise to bad conscience, the latter being a fact about human psychology. The "priestly reinterpretation of the animal's 'bad conscience' (cruelty turned backwards)" turned it into guilt and the feeling of "sin," the "most dangerous doom-laden feat of religious interpretation" in history (GM III:20). The reinterpretation of the psychological fact—internalized cruelty—tells the agent what that fact means, namely, that the agent is a sinner and should feel guilty. Yet, self-cruelty admits, obviously enough, of many other interpretations as to its meaning. *This case is clearly the paradigm case of "doing violence ... fabricating, falsifying" that Nietzsche has in mind.* It is the ascetic priest who, in this instance, is engaged in fabricating and falsifying; the facts about "animal psychology" are what they are, independent of the ascetic priest's imposition of meaning.

The *meaning* of the internalization of cruelty on Nietzsche's account is metaphysically constituted only from one affective perspective or another; in itself, the psychological phenomenon has no meaning. It is *not true*, after all, that humans are sinners, and we know Nietzsche does not think it is true: what is true is that *human beings suffer*, that *they have internalized their cruel instincts*, and that *"meaningless suffering is unbearable"* (GM III: 28), i.e., it leads to suicidal nihilism. The latter are claims in descriptive (if speculative) psychology/anthropology; it is only from particular perspectives (e.g., the ascetic priest's) that these psychological facts are assigned *meaning* or *value*. The way this operates in the case of the ascetic ideal is, of course, the central concern of the Third Treatise of GM, and only by understanding the ways in which it operates can we contribute, as GM III:12 has it, to knowledge about how human beings behave, and why they take ascetic moralities seriously, the central topic of GM III. Nietzsche here proceeds as Thucydides does: he describes how things seem from a particular evaluative perspective that he does not share, though unlike Thucydides, Nietzsche's polemical hostility to the perspective in question is rather more obvious.

In the Third Treatise, Nietzsche goes on to remark that the "ascetic priest has ruined the health of the soul wherever he has come to power" (GM III:22), though as the first line of the next section makes clear, it is a matter of ruining *Gesundheit und Geschmack* (health and taste; "health," of course, is a term of endorsement for Nietzsche, meaning one has a certain kind of "taste"). Thus, Nietzsche presents his own polemic against *The New Testament* as compared to the *Old* as involving him "standing alone in my taste regarding this most esteemed, most overestimated scriptural work (the taste of two millennia is *against* me)," yet adds that, notwithstanding, "I have the courage of my bad taste" (GM III:22). In other words, his claims about the *meaning* of the *Old* as against the *New Testament* reflect Nietzsche's *type* (with its admittedly "bad" taste) against the ascetic's *type*. *Different types have different tastes*, and as Zarathustra says, "all of life is a dispute about tastes" (Z II: 13).

But if all life is a dispute about tastes, and understandings of internalized cruelty vary by the type of person who experiences it, then someone, from the outside, trying to understand these phenomena will need to be able to represent all of these different affective perspectives in order to understand what is really going on. And that will require the inquirer to suspend his own opinions about the differing metaphysically constitutive perspectives on questions of meaning, given that those opinions will have been aroused in the first place by the perspectives he finds appealing. That ability to have "*in our power* our 'pros' and 'cons': so as to be able to engage and disengage them so that we can use the *difference* in perspectives and affective interpretations for knowledge" is the ability in which both Thucydides and (sometimes) Nietzsche excelled; as a result, they increased our knowledge of the phenomena they analyzed.

Conclusion

Nietzsche's perspectivism, based on the actual texts, is not equivalent to PaP, and in some ways is rather banal in the post-Quinean world. All knowing may depend on "will" or "affect," but evolutionary pressures select in favor of some of these affects, such that most "creatures like us" converge on many epistemic values, albeit not all. Yet, beyond that baseline, the Busy World Hypothesis reminds us that which particular objects of cognition command our attention will be influenced by our other affects and interests, and that this epistemic constitution plays an important role in what we know about the world.

Acknowledgments

Thanks to an audience at the University of Vienna conference on relativism in nineteenth- and twentieth-century German philosophy for useful discussion of the earliest draft. I am grateful to Chris Janaway for helpful comments on a later draft, and to Ken Gemes for comments on the most recent versions that forced a major, and I believe salutary, rethinking of the argument. Thanks to Martin Kusch and Johannes Steizinger for comments on the penultimate draft, and to Dan Telech for research assistance.

Abbreviations

References to Nietzsche are to the English-language acronyms for his books: *Daybreak* (D); *The Gay Science* (GS); *Beyond Good and Evil* (BGE); *On the Genealogy of Morality* (GM); *Thus Spoke Zarathustra* (Z); *Twilight of the Idols* (TI); *The Antichrist* (A); *The Will to Power* (WP). Roman numerals refer to parts or chapters, Arabic numerals to sections, not pages. I've consulted a variety of English translations (especially those by Walter Kaufmann, Judith Norman, and Maudemarie Clark and Alan Swensen), and occasionally made changes of my own based on the Colli-Montinari edition of the *Sämtliche Werke*.

Notes

1 There can be, of course, objective truths about relative or "relational" facts (cf. Leiter 2015a, 35–36, 85–90), but that is not what is at issue here.
2 I agree with Clark's devastating critique of Nietzsche's ill-formed views in *On Truth and Lie,* but I now think it is less important that his views on truth and knowledge evolved in precisely the way Clark originally argued.
3 Lange (1925, 41) asserts "the Sensationalism of Protagoras" is "combined with a relativity which may remind us of [German materialists] Büchner and Moleschott … It is not man in his universal and necessary qualities, but each individual in each single moment, that is the measure of things … Protagoras must be regarded wholly as a predecessor of the theoretical philosophy of Kant" (42).

4 "Truth," in particular, is discussed in many places by Nietzsche, but the points I made long ago (Leiter, 1994) about those passages still seem to me correct: none commits Nietzsche to PaP.

5 Nietzsche thus endorses a kind of "species relativism" when it comes to knowledge, thus marking him as on one side of the German "psychologism" debate. (The laws of thought are just laws of psychology, so relative to the particular agent.)

6 Nietzsche seems to have something like a Higher-Order-Thought (HOT) view of consciousness: cf. Riccardi (2017) for illuminating discussion.

7 Cf. Riccardi (2015, 225), which offers an illuminating discussion of why Nietzsche takes introspection to be an inadequate way to acquire self-knowledge, partly in the context of GS 354.

8 Cf. GS 112: "We operate only with things that do not exist: lines, planes, bodies, atoms divisible time spans, divisible spaces."

9 Nietzsche plainly thinks some of these errors are dispensable, e.g., about the good and about free will.

10 "*Das Perspektivische, die Grundbedingung alles Lebens.*" The contrast is made with Plato, who spoke *vom reinen Geiste und vom Guten an sich* (of pure spirit and the good in itself).

11 To be clear, Kant himself did not think that objectivity demanded knowledge of the noumenal world—for Kant, the objectivity of our knowledge *of the phenomenal world* was secured by the fact that the human mind necessarily imposed structures on phenomenal experience. But Nietzsche is following the post-Kantian skeptics who thought that the real upshot of transcendental idealism is that we are forever blocked from actual objective knowledge.

12 We can ignore "knowing" what the bottom of the table looks like based on testimony or other inferential sources of knowledge—Nietzsche makes clear with the reference to the eye in GM III:12 that perceptual knowledge is the relevant analogy.

13 More complicated affects (e.g., guilt vs. shame) may require cognitive content to individuate them, though whether their causal role in the explanation of behavior requires that is not clear: the cognitive component may turn out to be explanatorily otiose (cf. Leiter 2013, 241–246). It is telling that Nietzsche claims that *Triebe* (drives) are often unknown, but he never makes the same claim about *affects*.

14 On the *epistemic constitution* account, knowers are knowing more about the actual object, whose existence and character does not depend on the perspectives from which it is known. To be sure, the epistemic constitution view differs importantly from Kant, in the sense that it does not treat the object-in-itself as cognitively unavailable. Yet, from GS 354 and GS 374, we know that Nietzsche thinks that we can never get beyond the human perspective—"we cannot look around our own corner" (GS 374)—and thus "we cannot reject the possibility that the world *may include infinite interpretations*" but from perspectives ("corners") we can never occupy (GS 374). The upshot of Global Humeanism and Skeptical Darwinism is that there may be a world beyond our ken, but a world like that is of no practical relevance to human beings. That is consistent with the idea that GM III:12 involves only epistemic constitution: after all, we have already granted that Global Humeanism forms the backdrop for all knowing, including that which is the subject of GM III:12.

15 Ironically, Nietzsche was sometimes closer to Plato than Thucydides on this score!

16 Paul Woodruff's edition of Thucydides (1993, 89, 105).

References

Clark, M. (1990), *Nietzsche on Truth and Philosophy*, Cambridge: Cambridge University Press.

Goldie, P. (2002), *The Emotions: A Philosophical Exploration*, Oxford: Clarendon Press.

Janaway, C. (1997), *Beyond Selflessness*, Oxford: Oxford University Press.

Lange, F. A. (1925), *History of Materialism,* 3rd ed., London: Kegan Paul, Trench, Trubner & Co Ltd.

Leiter, B. (1994), "Perspectivism in Nietzsche's *Genealogy of Morals*," in R. Schacht (ed.), *Nietzsche, Genealogy, Morality,* Berkeley and Los Angeles: University of California Press, 334–357.

—— (2002), *Nietzsche on Morality*, London: Routledge.

—— (2013), "Moralities Are a Sign-Language of the Affects," *Social Philosophy & Policy* 30: 237–258.

—— (2015a), *Nietzsche on Morality*, 2nd ed., London: Routledge.

—— (2015b), "Normativity for Naturalists," *Philosophical Issues: A Supplement to Noûs* 25: 65–79.

Mann, J. E. and Getty L. L. (2011), "A Model Sophist: Nietzsche on Protagoras and Thucydides," *Journal of Nietzsche Studies* 42: 51–72.

Riccardi, M. (2015), "Inner Opacity: Nietzsche on Introspection and Agency," *Inquiry* 58: 221–243.

—— (2017), "Nietzsche on the Superficiality of Consciousness," in M. Dries (ed.), *Nietzsche on Consciousness and the Embedded Mind*, Berlin: De Gruyter, 93–112.

Stich, S. (1978), "Could Man Be an Irrational Animal?" in *Collected Papers, Volume 2: Knowledge, Rationality, and Morality, 1978–2010*, New York: Oxford University Press, 2012, 49–66.

Wilcox, J. T. (1974), *Truth and Value in Nietzsche: A Study of His Metaethics and Epistemology*, Ann Arbor: University of Michigan Press.

Woodruff, P. (1993), *Thucydides on Justice, Power and Human Nature*, Indianapolis, IN: Hackett.

9 Cassirer on relativism in science and morality

Samantha Matherne

Introduction

One of the central texts in which the German Neo-Kantian, Ernst Cassirer (1874–1945) addresses the topic of relativism is his often-overlooked work *Axel Hägerström* (1939). There, Cassirer engages with the theoretical, moral, and legal philosophy of Axel Hägerström (1868–1939), a leading member of the positivist movement known as "Scandinavian Realism" and whose work Cassirer had become familiar with during his years in exile in Sweden. Cassirer takes up the topic of relativism specifically in his discussion of Hägerström's moral philosophy, where he labels Hägerström's emotivist position "complete relativism" [völliger Relativismus] (Cassirer 1939, 54).[1]

Although Cassirer rejects Hägerström's complete version of relativism, in this chapter, I argue that Cassirer nevertheless endorses another form of relativism that is grounded on Kantian principles, which I shall label "critical relativism." On my reading, Cassirer thinks we need to embrace some form of relativism because our judgments cannot be absolutely true. However, unlike complete relativism, which he claims collapses into subjectivism and skepticism, Cassirer defends a critical form of relativism, which he thinks can still account for the objective validity of judgment.

In order to make his case against complete relativism and for critical relativism in *Axel Hägerström*, Cassirer considers how these issues bear not just on morality, but on natural science as well. Indeed, central to his argument in favor of critical relativism is the claim that our judgments in morality are like those in natural science: though they are not absolutely valid, they nevertheless can be objectively valid. For this reason, I shall focus on how Cassirer develops his account of relativism in the context of his analysis of the symmetry between judgments in natural science and morality. To be sure, a complete account of Cassirer's position on relativism would also have to take into account how this distinction between complete and critical relativism bears on his overarching philosophy of culture; however, this is not something I can pursue here.[2] Here, my more modest goal is to articulate Cassirer's framework for relativism and to analyze how he applies this framework to natural science and morality.

By my lights, pursuing his analysis of the parallel between natural science and morality promises to shed light not just on Cassirer's framework for relativism, but also on his theory of morality. This is significant because it is often called into question whether Cassirer has anything like a moral philosophy. Indeed, some readers have denied that Cassirer is concerned with morality at all. Leo Strauss famously notes and faults Cassirer on this count in his 1947 review of *The Myth of the State* (1946). There, Strauss objects that all Cassirer offers us are "inconclusive" remarks on the relationship between myth and politics, when what was really needed was "a radical transformation of the philosophy of symbolic forms into a teaching whose center is moral philosophy" (Strauss 1947, 128).[3]

More moderately, a number of commentators have claimed that although Cassirer does not offer an independent account of morality, his philosophy of culture has moral underpinnings.[4] To this end, interpreters draw on Cassirer's claims to the effect that, "human culture taken as a whole may be described as the process of man's progressive self-liberation," and that the progress of culture is the "progress of consciousness of freedom" (1944, 228; 1936, 90). Insofar as Cassirer thus treats freedom as the end of culture, these commentators argue that his philosophy of culture is ethical in orientation.

Although I am sympathetic to this ethical reading of Cassirer's philosophy of culture, in *Axel Hägerström* Cassirer seeks to defend a more robust moral philosophy. He makes this clear in the preface, where he writes:

> I have used the stimulus, which my study of Hägerström's major works has provided, to sharpen my own fundamental view [*Grundanschauung*], which I presented particularly in my *Philosophy of Symbolic Forms*, and to apply it to a new area. Thus my overall conception [*Gesamtauffassung*] of the problems of ethical and legal philosophy is treated much more extensively here than in my earlier writings, which especially concerned theoretical philosophy.
>
> (1939, 7)

Here Cassirer acknowledges that his major work on culture, his three-volume *Philosophy of Symbolic Forms* (1923, 1925, 1929b), was largely theoretical in orientation. And he expresses his intention to use *Axel Hägerström*, in part, as an opportunity to clarify his position on ethics and morality. This is why, in what ensues, Cassirer addresses questions surrounding "objective cognition [*Erkenntnis*]" and "*objective value-judgments*" in morality, the possibility of "*ethics as science* [*Wissenschaft*]," and the "phenomenology of ethical consciousness" (1939, 74, 59, 74). *Axel Hägerström* is thus a pivotal text for understanding Cassirer's moral philosophy.[5] And though I cannot offer a comprehensive overview of his moral thought here, I hope to make some headway in clarifying Cassirer's position by exploring his analysis of it in light of critical relativism and its parallel with natural science.[6]

I proceed as follows. In the first section I analyze Cassirer's account of the two kinds of relativism: complete and critical relativism. Then in section two, I consider how Cassirer applies this analysis of relativism to judgments in mathematical natural science. In the third section, I look at Cassirer's extension of this strategy to judgments in morality. My aim in sections two and three is to clarify Cassirer's claim that although judgments in natural science and morality are relative, they nevertheless can be objectively valid and are open to progress.

Two kinds of relativism

In *Axel Hägerström* and other writings, like *Substance and Function* (1910) and *Einstein's Theory of Relativity* (1921), Cassirer approaches the topic of relativism through a cognitive lens, as a theory about the truth or validity of judgment.[7]

At the most basic level, Cassirer takes relativism to be grounded in a negative commitment: it denies that our judgments can be *absolutely* true or valid.[8] Cassirer defines absolute truth or validity in terms of the idea of a judgment perfectly "copying" something absolute, i.e., something given to us as it is, independent from its relation to our minds or to other things (Cassirer 1921, 391).[9] Cassirer claims that, on the relativist view, absolute truth is beyond our grasp:

> To cognize [*erkennen*] the object [absolutely], our knowledge [*Wissen*] would, above all, have to be in a position to grasp it in its pure "in itself" and to separate it from all the determinations, which only belong to it relatively to us and other things. But this separation is impossible, not only actually, but in principle. For what is actually given to us only under certain definite conditions can never be made out logically as what it is in itself and under abstraction from precisely these conditions.
>
> (1921, 387, translation modified)

According to Cassirer, then, the basic relativist thought is that we cannot grasp things "in themselves" in the way required for absolutely true or valid judgments; we can only grasp them under "certain definite conditions." Our judgments are thus relative to a certain context, viz., these conditions and the things given to us in them. As a result, the relativist insists that we must recognize the "limit in principle which is set to all cognition and by which it is separated once for all from the definite apprehension of the truth as 'absolute'" (1921, 387, translation modified).

However, Cassirer indicates that there are different kinds of relativism that can be built on the basis of this negative thesis, depending on which theory of *objective* truth is assumed. In particular, Cassirer draws a distinction between *complete relativism*, which rests on an absolute theory of objective truth, and, what I am calling *critical relativism*, which rests on a critical, i.e., Kantian, theory of objective truth. As we shall see, Cassirer thinks that this basic commitment regarding objective truth gives rise to relativisms that are very different in character, the former being one he rejects, and the latter being one he endorses.

Let us begin with complete relativism. According to Cassirer, the complete relativist is committed to the view that a judgment can be objectively true or valid only if it is absolutely true or valid, i.e., only if it corresponds to something absolute. Since the complete relativist also denies that we can make absolutely true judgments, she thinks our judgments cannot be objectively true. Instead, she claims that our judgments are merely subjectively valid: they are only "true to" the individual who makes them (1921, 393).[10] Given this view of the subjective limitations of judgment, Cassirer claims that complete relativism ultimately results in skepticism.[11]

In the third chapter of *Axel Hägerström*, Cassirer identifies Hägerström's account of morality as an instance of complete relativism. In the background of this suggestion is Cassirer's earlier analysis of Hägerström's theory of reality and objective truth in the first two chapters.[12] As Cassirer points out there, on the one hand, Hägerström is an anti-metaphysical thinker who denies that there exists anything like a thing-in-itself that underlies the empirical reality that is given to us.[13] On the other hand, Hägerström rejects the "subjectivist" thesis that the only thing immediately given to us is consciousness.[14] Instead, Hägerström argues that what is absolute is an empirical reality that is "absolutely independent from the thinking subject" and that is "presupposed" by all our judgments (1939, 19, 48). He characterizes this reality as something that is "immediate," "unconditioned," "self-identical," and characterized by "determinateness" (1939, 13–14, 19–20, 50). He also maintains that this reality sets the standard of objective truth or validity: a judgment is objectively valid if it "apprehends something as real," i.e., if it apprehends the reality that is immediately given to us (1939, 48–49).

According to Hägerström, although it is possible to make objectively true judgments in natural science, we cannot make them in morality. Indeed, as Cassirer emphasizes, Hägerström advocates "subjectivism" when it comes to morality, claiming that "the search for any objectivity [in morality] is a misguided ... endeavor" (1939, 53). On Cassirer's reconstruction, Hägerström takes this to be the case because he believes that, in morality,

> what we find are ... merely "valuations" [*Bewertungen*], which as pure expressions of feeling [*Gefühlsausdrücke*] are bound to the feeling and desiring I, and which can therefore never go beyond the circle of the individual [*Kreis des Individuellen*], which is valid for this or that subject.
>
> (1939, 53)

Hägerström thus defends an emotivist theory of morality, according to which all moral judgments are mere expressions of feeling and, as such, are valid only for the subject who makes them. Insofar as all moral judgments are equal in their subjective validity, Hägerström treats them all as "equivalent." As Cassirer points out, Hägerström then draws the skeptical conclusion that "ethics as science" and "objective value judgments" are not possible. And it is for these reasons that Cassirer labels Hägerström's moral philosophy "complete relativism" (1939, 54, 59).

There is, however, another form of relativism that Cassirer discusses, which I am calling "critical relativism" because it is grounded in critical, i.e., Kantian, principles.[15] To be sure, as a form of relativism, critical relativism denies the possibility of us making absolutely true or valid judgments. And in critical relativism, Cassirer indicates that the motivation for this denial stems from the Kantian recognition of the "relativity of cognition," i.e., the recognition that cognition is restricted to the realm of appearances or phenomena and that cognition of things-in-themselves or noumena is off-limits (1921, 392, translation modified). As Cassirer articulates this Kantian insight, "we never know [*kennen*] things as they are for themselves, but only in their mutual relations" (1910, 305). Insofar as the Kantian thus rejects the possibility of us making judgments that correspond to things-in-themselves, Cassirer indicates she is committed to some version of relativism.[16]

However, Cassirer argues that in spite of the relative nature of cognition, the Kantian also has a theory of objective truth, which allows her to account for the possibility of objectively true or valid judgments. Cassirer labels this Kantian theory of truth the "critical" or "idealistic" theory of truth (1910, 319, 391).[17] And he asserts that this theory

> does not measure the truth of fundamental cognitions by transcendent objects, but it grounds conversely the meaning of the concept of the object on the meaning of the concept of truth. Only the idealistic concept of truth overcomes finally the conception which makes cognizing [*Erkennen*] a copying … The "truth" of cognition changes from a mere pictorial expression [*Bildausdruck*] to a pure functional expression [*Funktionsausdruck*].
>
> (1921, 391, translation modified)

As we see in this passage, Cassirer suggests that the critical concept of truth measures objective validity not in terms of whether a judgment "copies" a thing in itself, but rather in terms of whether that judgment reflects the "functions," i.e., the concepts, laws, or principles, that the objects of cognition (appearances, phenomena) conform to.

Although our judgments are relative to the context of cognition, Cassirer argues that the Kantian does not conceive of this context in merely subjective terms. To begin, he claims that the Kantian regards these functions as those that belong to thinking or understanding in general and thus transcend any individual's mind.[18] Moreover, he maintains that, on the critical view, the objects that these functions condition are "transcendent" from the standpoint of the psychological individual" (1910, 297). For these reasons, he asserts that the critical view "recognizes no other and no higher objectivity than that, which is given in experience itself and according to its conditions" (1910, 278).[19] And the Kantian, Cassirer maintains, identifies this objective context of cognition as the one that our judgments are relative to.

Insofar as the context of cognition is thus an objective one, Cassirer argues that it provides the Kantian with an alternative way to account for objective validity within a relativist framework. She sets the objective context or "whole" of cognition as the standard by means of which we measure a judgment (1910, 277, 284; 1995, 117).[20] A judgment is thus objectively true or valid if it expresses the functions that govern this whole and that condition the objects in it.

However, Cassirer offers his own modification of this critical theory of truth, emphasizing the possibility of our judgments having different "degrees" of objective validity (1910, 277). Although we measure the objective validity of our judgments against the whole of cognition, Cassirer argues that each judgment nevertheless has a specific "sphere of validity," i.e., a sphere of objects for which it holds (1910, 278). He also maintains that we can then measure the "degree of objectivity" that each judgment has by determining how broad that sphere of validity is (1910, 277). A judgment that is valid with respect to a broader sphere of objects of cognition will have a higher degree of objective validity than a judgment that has a narrower sphere of validity. And a judgment with the highest degree of validity would be one that articulates the functions that govern the whole of cognition in a unified and universal way, i.e., one that expresses those functions that govern all possible objects of cognition in all possible circumstances.[21] On Cassirer's view, then, objective validity is not an all-or-nothing thing; rather, it is something that comes in degrees. As we shall see below, Cassirer thinks that this is crucial for being able to account for the progress of our judgments as we move from judgments that are less true, i.e., that have narrower spheres of objective validity, to those that are more true, i.e., that have broader spheres of objective validity.

Unlike complete relativism, then, Cassirer thinks that critical relativism is able, on the one hand, to acknowledge that our judgments cannot be absolutely valid, while still accounting for their objective validity, on the other. Let us now turn to how Cassirer makes use of this view of critical relativism to account for the objective validity and progress of judgments in natural science (section two) and morality (section three).

Critical relativism and natural science

I want to begin by considering why Cassirer thinks that relativism of any form is called for in order to account for our judgments in natural science. Recall that, on his view, absolute truth requires making a judgment that corresponds to a thing-in-itself, and that this is, in principle, impossible because the only objects we can cognize are phenomena that are relative to the conditions of cognition. Extending this line of thinking to the case of science, Cassirer writes,

> we determine the object not as an absolute substance beyond all cognition, but as the object shaped in progressing experience ... This object may be called "transcendent" from the standpoint of a psychological individual; from the standpoint of logic and its supreme principles, nevertheless it is to be characterized as purely immanent. It remains strictly within the sphere, which these principles determine and limit, especially the universal principles of mathematical and scientific cognition ... There results, strictly speaking, no absolute, but only relative being.
>
> (1910, 297–298, translation modified)

As we see in this passage, Cassirer analyzes the "relative being" of objects not in terms of dependence on the mind of a "psychological individual" but rather in terms of dependence on "progressing experience." He is here defining "experience" in the technical way that the Marburg Neo-Kantians do, viz. as the experience involved in mathematical natural science and which is governed by the principles of logic, mathematics, and natural science.[22] On Cassirer's view, then, the objects we cognize in natural science are ones that are conditioned by the progress of natural science. Given that these objects thus depend on the progress of scientific experience, Cassirer concludes that the judgments of natural science cannot be absolutely true. Rather, Cassirer claims that in science "we face ... a perpetually self-renewing process with only relative stopping-points; and it is these stopping-points, which define the concept of 'objectivity' at any time" (1910, 278).

At this point, however, Cassirer does not proceed as the complete relativist does, i.e., by drawing the skeptical conclusion that since our judgments in natural science cannot be absolutely valid, they cannot be objectively valid. Instead, he proceeds in the manner of critical relativism, arguing that the objective validity of natural science is possible relative to the whole of scientific cognition. Cassirer typically refers to this whole as the "whole of experience" or as "nature"; however, on Cassirer's Kantian view, nature just is what conforms to the conditions of experience, so he conceives of these wholes as interchangeable.[23] And he argues that the whole of experience/nature is the standard by means of which we can measure the objective validity of our scientific judgments:

> we do not measure presentations with respect to absolute objects ... Each partial experience [*Teilerfahrung*] is ... examined as to what it means for the total system; and this meaning determines its degree of objectivity. In the last analysis, we are not concerned with what a definite experience "is," but with what it "is worth," i.e., with what function it has as a particular building-stone in the structure of the whole.
>
> (1910, 277)

According to Cassirer, then, we can determine the objective validity of a judgment in natural science by measuring it against the whole of experience/nature: if that judgment articulates the functions of experience that condition physical objects, then it is objectively valid. Thus, for Cassirer, the whole of experience/nature serves as his alternative to the standard of absolute truth.

Moreover, continuing on the trajectory we discussed above, Cassirer claims that the whole of experience/nature allows us to determine the degree of objectivity of our judgments in natural science. For, on Cassirer's view, one theory will be "truer" or more objective than another if it can accommodate a "larger sphere of experience" (1939, 67). If we compare, for example, classical mechanics and special relativity, Cassirer claims that special relativity has a higher degree of objectivity than classical mechanics because it provides us with a more systematic account of physical phenomena across inertial reference frames (1921, Ch. 2).

Furthermore, Cassirer maintains that we can use these considerations about the degree of objectivity of scientific judgments to measure the progress of natural science.[24] On his view, natural science progressively converges toward an "ideal limit":

> Knowledge realizes itself only in a succession of logical acts, in a series that must be run through successively, so that we may become aware of the rule of its progress. But if this series is to be grasped as unity, as an expression of an identical reality, which is defined the more exactly the further we advance, then we must conceive the series as converging towards an ideal limit. This limit "is" and exists in definite determinateness, although for us it is not attainable save by means of the particular members of the series and their change according to law.
>
> (1910, 315)

The ideal limit, then, would be a theory that articulates the functions that govern the whole of experience/nature in a unified and universal way.[25] It is in virtue of converging toward this ideal limit that Cassirer believes that natural science advances. Framing this conception of progress in terms of objective validity, we find that, on Cassirer's view, the progress of natural science involves advancing toward judgments whose objective validity extends over a broader sphere of possible scientific experience.

Thus, although Cassirer thinks we must acknowledge that scientific judgments are relative because they cannot be absolutely valid, he does not think this means they are merely subjective. Rather, he sets up an alternative standard of objective truth, that of the whole of experience/nature, which he thinks can account for the objective validity and progress of judgments in natural science. As we now turn to Cassirer's analysis of morality, I aim to show that he uses critical relativism in a similar way to demonstrate the possibility of objectivity and progress in the moral realm.

Critical relativism and morality

In *Axel Hägerström*, Cassirer orients his account of morality around the parallels he sees between it and natural science. He does so because he thinks that if we accept that objectively valid judgments and progress are possible in natural science, and we accept that there are important affinities between judgments in natural science and morality, then we will accept that morality can involve objectively valid judgments and progress as well.

In order to tease out this parallel, Cassirer indicates that judgments in natural science and moral judgments do not copy some absolute thing-in-itself. Rather, they refer to a realm of phenomena. Whereas scientific judgments refer to the realm of "nature," he claims that moral judgments refer to the "realm of willing" [*Wollen*] and volitional phenomena, e.g., choices or actions. And he argues that the phenomena in both realms are governed by functions: just as the physical phenomena in nature are governed by scientific concepts, laws, and principles, so too does he believe that volitional phenomena are governed by "rules for willing and action" (1939, 71). He takes this to be the case in morality because he thinks that moral judgments involve a consciousness of the "justification" of our actions, i.e., a sense of whether our actions were ones we should have done or not (1939, 63). According to Cassirer, this reveals that moral judgments have the logical form of "subsumption": they subsume a particular action under a general principle (1939, 63). This being the case, far from our moral judgments simply expressing feeling as Hägerström would have it, Cassirer maintains that they articulate the functions that govern volitional phenomena, just as our scientific judgments articulate the functions that govern physical phenomena.

Given that, on Cassirer's view, the judgments in both realms are concerned not with absolute things in themselves, but rather with phenomena that are determined by rules and functions, our judgments are not absolutely valid. Nevertheless, Cassirer endeavors to show that, *pace* Hägerström, this relativity is as consistent with the objective validity of our moral judgments as it is with the objective validity of our judgments in natural science.

In order to establish this point, Cassirer argues that in morality we find the same sort of alternative measure that allows us to account for the objective validity of its judgments as we did in natural science. More specifically, he claims that, like with the whole of experience in science, we can think of the whole of willing, or what he sometimes calls the "unity of willing," as the "standard" [*Maßstab*] by means of which we measure the objective validity of our moral judgments (1939, 69, 75). On this view, a moral judgment is objectively valid if it articulates the principles that govern the whole of action and willing.

Now, as a Kantian, Cassirer does not think that ascertaining which principles govern moral judgments is an empirical matter; rather, he takes it to be a normative matter, a matter of articulating the rules that *ought* to govern choices and actions. Moreover, he claims that these principles are to have "unity and universality," i.e., they are supposed to govern all possible actions and choices in the realm of willing (1939, 74). For these reasons, Cassirer takes the rules that govern the realm of willing and action to be ultimately connected to the "concept of duty" and the obligations that hold in an unexceptional, unified, and universal way for all choices and actions. Like Kant, Cassirer conceives of the rules of duty as rules that govern autonomous actions, i.e., actions that arise through an "active self-binding" in which we act in accordance with laws we give to ourselves (1939, 103–104). For Cassirer, then, a moral judgment is objectively valid if it expresses the rules that govern an autonomous will bound by duty and hold in a unified and universal way for all possible volitional phenomena.

Then making a move similar to the one he made in his discussion of natural science, Cassirer argues that we can use the whole of willing to measure both the degree of objective validity of moral judgments and progress in morality. For, on Cassirer's view, a moral judgment will have a higher degree of objective validity if it rests on principles that govern the realm of willing in a more universal way, and a lower degree of objective validity if its principles are restricted to specific circumstances. Indeed, Cassirer claims that this standard can operate as a "principle of super- and sub-ordination" of different moral judgments, marking judgments that are more advanced than others (1939, 69, 67).

Cassirer, in turn, maintains that this standard puts us in a position to trace a "phenomenology of ethical consciousness," as morality progresses toward theories that are more universal (1939, 79).[26] Making this point in Kantian terms, Cassirer claims that we can trace the progress in morality away from theories that turn on "hypothetical rules" that are contingent on particular ends and circumstances, to those that rest on "categorical imperatives," i.e., rules that express the moral duty of an autonomous will (1939, 67). He identifies judgments made according to the principles of Stoic and Kantian ethics as examples of the more advanced kind of moral theories (1939, 79).

According to Cassirer, what these considerations ultimately point toward is the "scientific character" of morality (1939, 62). For, on his view, "objective cognition" is as possible in morality as it is in natural science: not only is morality capable of issuing in objectively valid judgments, but also it involves progress toward theories with higher degrees of objective validity that are more universal in character (1939, 74). Thus, far from morality collapsing into complete relativism, Cassirer uses the framework of critical relativism to show that it is at once relative and objective.

Conclusion

My aim in this chapter was to clarify Cassirer's distinction between complete and critical relativism and to analyze how he uses the framework of critical relativism to account for the possibility of objectively valid judgments and progress in both natural science and morality. As I indicated in the introduction, I regard Cassirer's discussion of the objectivity and progress of morality in *Axel Hägerström* to be particularly significant because it has often been called into question whether Cassirer has anything approaching a moral philosophy. Although I could by no means offer an exhaustive analysis of Cassirer's moral philosophy here, I hope to have made some headway on this issue by bringing to light the parallel he draws between the objectivity and progress of natural science and morality in the framework of critical relativism.

Acknowledgments

I would like to thank Kristin Gjesdal, Katherina Kinzel, Martin Kusch, Michael Morris, Hans-Christoph Schmidt, Sebastian Stein, Johannes Steizinger, Chris Yeomans, Paul Ziche, Günter Zöller, and audiences at the *History of Relativism* Conference in Vienna in 2016 and the *Text and Context* Conference in Munich in 2017 for helpful feedback on earlier versions of this chapter.

Notes

1 Translations of *Axel Hägerström* are my own.
2 See Luft (2005, 13–14; 2015, 209–221) and Freudenthal (1996, 2008) for the argument that in his philosophy of culture, Cassirer endorses pluralism rather than relativism. See Rudolph (2008) for a discussion of Cassirer's view of culture in light of Blumenberg's reading of it as a kind of historical relativism.
3 Strauss continues by saying that a moral transformation of the philosophy of symbolic forms would involve "a return to Cassirer's teacher Hermann Cohen, if not to Kant himself" (Strauss (1947, 128). Making a similar point elsewhere, Strauss claims, "having been a disciple of Hermann Cohen [Cassirer] had transformed Cohen's philosophic system, the very center of which was ethics, into a philosophy of symbolic forms in which ethics has silently disappeared" (Strauss 1988, 246). For a discussion of Strauss's relationship to Cassirer (and Heidegger), see Gordon (2010, 315–319, 343–345).
4 For a discussion of the ethical dimensions of Cassirer's philosophy of culture, see Lofts (2000, 202–206), Habermas (2001), Skidelsky (2008, 103–105), Gordon (2010, 360–362), Kreis (2010, Ch. 15), Verene (2011, 95–96, 109–111) and Luft (2015, 221–231).
5 In addition to the texts I discuss in this essay, Cassirer addresses issues pertaining to moral philosophy in *Freiheit und Form* (*Freedom and Form*) (1916), *"Die Idee der republikanischen Verfassung"* (*The Idea of the Republican Constitution*) (1929a), *Vom Wesen und Werden des Naturrechts* (*On the Essence and Becoming of Natural Law*) (1932), and Ch. 13 of Cassirer 1956, "Concluding Remarks and Implications for Ethics."
6 For a more extended discussion of Cassirer's moral philosophy, see Krois (1987, Ch. 4), Coskun (2000), Matherne (2019, Ch. 7).
7 He addresses the issue of relativism in these cognitive terms at (1910, 187) and in (1921, Ch. 3). Insofar as his analysis of relativism in these passages is cognitive in

orientation, his approach differs from more contemporary approaches that focus on the issue of cultural relativism. See Freudenthal (1996) for the argument that Cassirer's relativism extends only to his account of cognition, not to his account of culture.

8 For Cassirer's discussion of the notion of absolute truth, see Cassirer (1910, 187, 1921, Ch. 3 and 1995, 116–117).

9 In discussing different historical forms of relativism, Cassirer argues that in the ancient world, the relevant absolute is an "absolute thing" outside of us and in the empiricism of the early modern era it is an "absolute sensation" (see 1921, 390–391).

10 In *Einstein's Theory of Relativity* Cassirer discusses this version of relativism under the heading of "relativistic positivism," and there targets Petzoldt (1912).

11 In the third chapter of *Einstein's Theory of Relativity*, Cassirer discusses three such forms of skepticism: ancient skepticism, Humean skepticism, and skepticism in light of the theory of relativity.

12 For Hägerström's own summary of his theoretical and moral philosophy, see Hägerström (2002, 33–72, 313–315).

13 See Cassirer's summary of Hägerström's anti-metaphysical position in *Axel Hägerström*, Ch. 1, "The Fight Against Metaphysics."

14 See Cassirer's summary of Hägerström's rejection of subjectivism in *Axel Hägerström*, Ch. 2, "The Critique of Subjectivism."

15 In addition to his discussion of critical relativism in *Axel Hägerström*, see also Cassirer (1921, 391–393 and 1995, 117, 120).

16 See, e.g., his discussion of relativism in "On Basis Phenomena," where he describes the "critical" position as a "relative" position (117). See also his discussion of Kant's commitment to relativism in *Einstein's Theory of Relativity*, where he says: "Kant ... gained his own interpretation of the critical concept of the object, in which the relativity of cognition was affirmed in a far more inclusive meaning than in ancient or modern skepticism, but in which also this relativity was given a positive interpretation" (Cassirer 1921, 392, translation modified).

17 In "On Basis Phenomena" Cassirer also calls this view of truth "relative truth" (1995, 117).

18 This is symptomatic of Cassirer's anti-psychologistic reading of these functions. See, e.g., his claim that, these principles depend "not on any concrete psychic contents or acts" and that, "to the psychological immanence of impressions is opposed ... the logical universality of the supreme principles of cognition [*Erkenntnisprinzipien*]" (Cassirer 1910, 300, translation modified). See Matherne (2018) for a more extended discussion of Cassirer's account of these functions and their relation to psychology.

19 Although in this passage, Cassirer has in mind the sort of "experience" involved in science, we shall see below that he extends this account of objectivity to morality.

20 I will specify below what the relevant wholes for natural science and morality are.

21 Cassirer often labels "invariants" the functions that govern the whole of cognition in a unified and universal way (see, e.g., Cassirer 1910, 249–250, 268–270, 273).

22 See, e.g., Cohen, "experience, which is mathematical natural science" (Cohen 1871/1885, 501, my translation). For a discussion of the Marburg conception of experience, see Richardson (2003) and Holzhey (2005).

23 He describes it as the whole of experience at (1910, 278, 284–285, 301–302) and (1995, 117) and as nature at (1921, 345, 374, 416–417) and (1939, 71).

24 For a more extended view of Cassirer's account of progress in science and its contrast with the Kuhnian account, see Friedman (2005) and (2008).

25 Cassirer sometimes makes this point by drawing on Planck's notion of a unified
physical theory, which has "unity in respect of all features of the [physical] picture,
unity in respect of all places and times, unity with regard to all investigators, all
nations, all civilizations" (Cassirer 1910, 307, quoting Planck (1909).

26 Although I cannot explore this further here, Cassirer does not think that this progress is unassailable. Rather, as he makes clear in *The Myth of the State* (1946),
moral regress not only is possible, but it actually occurred during the rise of
fascism in the twentieth century.

References

Cassirer, E. (1910), *Substance and Function*, translated by W. Swabey and M. Swabey,
Chicago: Open Court, 1923.

—— (1916), *Freiheit und Form. Studien zur deutschen Geistesgeschichte*, Berlin: Bruno
Cassirer.

—— (1921), *Einstein's Theory of Relativity,* translated by W. Swabey and M. Swabey,
Chicago: Open Court, 1923.

—— (1929a), *Die Idee der Republikanischen Verfassung: Rede zur Verfassungsfeier am
11. August 1928*, Hamburg: Friederichsen.

—— (1929b), *The Philosophy of Symbolic Forms. Volume Three: The Phenomenology
of Knowledge*, translated by R. Manheim, New Haven: Yale University Press, 1957.

—— (1932), *Vom Wesen und Werden des Naturrechts*, Berlin: Felix Cassirer.

—— (1936), "Critical Idealism as a Philosophy of Culture," in *Symbol, Myth, and
Culture: Essays and Lectures of Ernst Cassirer: 1935–1945*, edited by D. Verene,
New Haven: Yale University Press, 1981, 64–91.

—— (1939), *Axel Hägerström: Eine Studie zur Schwedischen Philosophie der Gegenwart*,
Göteborg: Göteborgs Högskolas Årsskrift 45, reprinted as *Gesammelte Werke
vol. 21*, edited by B. Recki, Hamburg: Meiner Verlag, 2005.

—— (1944), *An Essay on Man*, New Haven: Yale University Press.

—— (1946), *The Myth of the State*, New Haven: Yale University Press.

—— (1956), *Determinism and Indeterminism in Modern Physics: Historical and
Systematic Studies of the Problem of Causality*, translated by O. T. Benfey, New
Haven: Yale University Press.

—— (1995), "On Basis Phenomena," in *The Philosophy of Symbolic Forms, Vol. 4,
The Metaphysics of Symbolic Forms*, translated by J. M. Krois, New Haven: Yale
University Press, 115–192.

Cohen, H. (1871/1885), *Kants Theorie der Erfahrung*, Berlin: F. Dümmler.

Coskun, D. (2007), *Law as Symbolic Form: Ernst Cassirer and the Anthropocentric
View of Law, Law and Philosophy Library* 82, Dordrecht: Springer.

Friedman, M. (2005), "Ernst Cassirer and the Philosophy of Science," in *Continental
Philosophy of Science*, edited by G. Gutting, Malden: Blackwell, 69–83.

—— (2008), "Ernst Cassirer and Thomas Kuhn: The Neo-Kantian Tradition in
History and Philosophy of Science," *The Philosophical Forum* 39(2): 239–52.

Freudenthal, G. (1996), "Pluralism or Relativism? I. Yehuda Elkana's 'Two-Tier
Thinking'," *Science in Context* 9(1): 151–163.

—— (2008), "The Hero of the Enlightenment," in *The Symbolic Construction of
Reality: The Legacy of Ernst Cassirer*, edited by J. Barash, Chicago: The University
of Chicago Press, 189–213.

Gordon, P. (2010), *Continental Divide: Heidegger, Cassirer, Davos*, Cambridge, Mass.: Harvard University Press.

Habermas, J. (2001), *The Liberating Power of Symbols: Philosophical Essays*, translated by P. Dews, Cambridge, Mass.: MIT Press.

Hägerström, A. (2002), *Philosophy and Religion*, Oxford: Routledge.

Holzhey, H. (2005), "Cohen and the Marburg School in Context," in *Hermann Cohen's Critical Idealism,* edited by R. Munk, Dordrecht: Springer, 3–37.

Kreis, G. (2010), *Cassirer und die Formen des Geistes*, Berlin: Suhrkamp.

Krois, J. M. (1987), *Cassirer: Symbolic Forms and History*, New Haven: Yale University Press.

Lofts, S. (2000), *Ernst Cassirer: A "Repetition" of Modernity*, Albany: State University of New York Press.

Luft, S. (2005), "Cassirer's Philosophy of Symbolic Forms: Between Reason and Relativism: A Critical Appraisal," *Idealistic Studies* 34(1): 25–47.

—— (2015), *The Space of Culture: Towards a Neo-Kantian Philosophy of Culture (Cohen, Natorp, and Cassirer)*, Oxford: Oxford University Press.

Matherne, S. (2018), "Cassirer's Psychology of Relations: From the Psychology of Mathematics and Natural Science to the Psychology of Culture," *Journal for the History of Analytic Philosophy,* Special Issue: Method, Science, and Mathematics: Neo-Kantianism and Analytic Philosophy 6(1): 132–162.

—— (2019), *Cassirer*, London: Routledge.

Petzoldt, J. (1912), "Die Relativitätstheorie im erkenntnistheoretischen Zusammenhang des relativistischen Positivismus," in *Deutsche Physikalische Gesellschaft, Verhandlungen* 14: 1055–1064.

Planck, M. (1909), *Die Einheit des physikalischen Weltbildes*, Leipzig: Hirzel.

Richardson, A. (2003), "Conceiving, Experiencing, and Conceiving Experiencing: Neo-Kantianism and the History of the Concept of Experience," *Topoi* 22(1): 55–67.

Rudolph, E. (2008), "Symbol and History: Ernst Cassirer's Critique of the Philosophy of History," in *The Symbolic Construction of Reality: The Legacy of Ernst Cassirer*, edited by J. Barash, Chicago: The University of Chicago Press, 3–16.

Skidelsky, E. (2008), *Ernst Cassirer: The Last Philosopher of Culture*, Princeton: Princeton University Press.

Strauss, L. (1947), "Review of the Myth of the State by Ernst Cassirer," *Social Research* 14(1): 125–128.

—— (1988), *What Is Political Philosophy? And Other Studies*, Chicago: University of Chicago Press.

Verene, D. (2011), *The Origins of the Philosophy of Symbolic Forms: Kant, Hegel, and Cassirer*, Evanston: Northwestern University Press.

10 Simmel and Mannheim on the sociology of philosophy, historicism, and relativism

Martin Kusch

Introduction

This chapter identifies and evaluates some central relativistic and historicist themes in early-twentieth-century German sociology of knowledge. The two main figures are Georg Simmel (1858–1918) and Karl Mannheim (1893–1947).

Concentrating on Simmel and Mannheim might seem surprising in a volume focused on *philosophical* positions concerning relativism. Indeed, Simmel and Mannheim are today primarily remembered as among the most influential "founding fathers" of the social sciences. It is important, however, to remember that the two men worked in an academic world in which the borders between philosophy and the social sciences were unclear and open (cf. Goodstein 2017). Moreover, Simmel never held a chair in sociology and did not think of himself as a sociologist. And both thinkers put forward ideas and theories that were recognized as philosophical *and* sociological by their contemporaries.

Neither "relativism" nor "historicism" had a precise and agreed-upon meaning in early-twentieth-century debates. Still, put in very general terms, "relativism" referred to the denial of truths that are universal, fixed, and independent of human psychology and cultures. "Historicism" meant the idea that, to understand any phenomenon, one needs to understand its history. Obviously, these two ideas could be combined.

Simmel and Mannheim systematically reflected on the relationship between relativism and historicism. They were not the first to do so. Historical and philosophical theorizing about the status of history (as a field of study) had already highlighted the tensions between historical contingency and philosophical normativity. The novel element in Simmel and Mannheim was their effort to bring these issues to bear on the philosophy of the social sciences in general, and on the sociology of knowledge in particular.

In Simmel's and Mannheim's work, relativism and historicism featured in at least three different ways. First, both developed and defended distinctive general claims about various kinds of "relativism." Second, in their historical case studies, Simmel and Mannheim made use of what they regarded as, "relativistic" methodologies. And third, both made the emergence of, relativism and historicism a central topic for their sociological-historical inquiries. Finally, to these three dimensions of relativism *within* Simmel's and Mannheim's writings, we can add a fourth that goes *beyond*: their writings triggered vigorous and extensive debates about relativism.

My overall evaluative thesis is that my early sociologists of knowledge were not fully successful in dealing with relativism. Simmel declared himself a relativist, but it remains unclear what precisely his relativism amounted to. Mannheim had a clearer view of relativism but thought of it as something the sociology of knowledge had to avoid at all costs. But he hardly succeeded in his attempts to do so. Fortunately, in neither case did the lack of success regarding the philosophical handling of relativism weaken the quality and interest of the sociological-historical case studies.

Simmel on money and philosophy

One natural starting point for my investigation is Simmel's *Philosophie des Geldes* (*Philosophy of Money*, 1900). I am here only considering its nascent sociology of philosophy, to wit, its attempts to identify parallels between philosophical ideas and economic practices or theorizing. Simmel claimed that such type of inquiry had been made possible by the "worldview" of "relativism" (1900, 142).

The central contrast running through the book was between modern and ancient/medieval interpretations of money. The ancient or medieval rendering amounted to a "materialist conception": the value of money is the "substance" out of which it is made. Moreover, influential thinkers, such as Aquinas, insisted that money should not be traded or lent for interest. Closely related to this was the "just price doctrine," the idea of "a direct relationship between object and money-price." As Simmel had it, the substance-value conception was "the appropriate theoretical expression of an actual sociological condition," and the conception fell apart when the condition disappeared. The condition in question was that people in ancient and medieval times valued "landed property" and "an agrarian economy ... with few and hardly variable intermediaries" (1900, 125, 167–168, 173, 235).

This rough sketch suffices for us to see how Simmel linked the "materialist conception of money" to philosophy. He proposed that this conception was inseparable from—that is, both supported by and supportive of—the metaphysical *topoi* of fixed substances with essential or accidental properties; a static, eternal cosmos; a static social organization; and forms of philosophy focused on qualitative rather than quantitative categories (1900, 148, 168, 234).

Turning from antiquity and the Middle Ages to modernity, Simmel insisted that the modern, "transcendental," understanding of money radically broke with its predecessors. It was Adam Smith who first articulated the new "transcendental theory." Most human relationships were now understood as relationships of exchange. Indeed, economic exchange was rendered the basic building block of society. Exchange was taken to be "as productive and value-creating as is production itself." Simmel emphasized especially the ways in which the market mechanism turns numerous different subjective preferences into one objective value (i.e., the stable price) (1900, 75–76, 81, 99, 173).

Simmel found numerous parallels between key aspects of modern monetary reality or economic theorizing and philosophical reflection. Suffice it here to mention three. The first concerned "relativity": the economic value of each good was dependent upon—and thus relative to—the economic value of every other good. Money "symbolized" such relativity. The relativity of money had its philosophical analogue in the emergence of various forms of relativism. The second parallel might be called the "dual perspective point": on the one hand, money stood aside from, or opposite, all goods that could be bought and sold with its help. But, on the other hand, for the moderns, money was also itself a good. In the philosophical domain this corresponded to a view of epistemic norms as both constitutive of space and time, and as empirical-psychological phenomena in spatio-temporally situated subjects, and a view of the ego as *both* transcendental *and* empirical. The third key link between modern money and philosophy concerned "condensation." For the modern thinkers, money "condensed" the value of things. The philosophical counterparts of this idea varied: they included the notion of "laws of nature" as condensations of "endless particular cases"; the modern idea of the "state" that—with the help of the civil service—functioned as a condensation of political powers; and the conception of "objectivity" as condensed intersubjectivity (1900, 79, 118–119, 128, 197, 303).

Simmel on historicism and relativism

Simmel emphasized repeatedly that the modern understanding of money went together with philosophical relativism and historicism. This naturally invites the question of how Simmel conceived of relativism and historicism. Simmel's most influential intervention into the historicism debates—the *Probleme der Geschichtsphilosophie* (*Problems of the Philosophy of History*, 1892)—discussed the "historical *apriori*." (Simmel was not the first to put forward this idea, but he developed it in exceptional detail.) This is the idea that the historian invariably imputes mental states to historical actors, and that the selection of such states depends crucially on the personality and social background of the historian in question. Accordingly, Simmel rejected Leopold von Ranke's (1795–1886) demand according to which historians must "eliminate themselves" in order to understand the past. Moreover, Simmel took his insight as confirmed by the fact that there is space for a sociological study of historical scholarship: it is usually possible to identify the social position of the historians from their texts. In some places, Simmel accepted the natural implication that all historical work is invariably perspectival (1892, 304, 321, 325, 328).

Simmel's "historical *apriori*" resonated with many other authors at the time. One obvious case in point was Theodor Lessing's (1872–1933) *Geschichte als Sinngebung des Sinnlosen* (*History as the Imputation of Sense to the Senseless,* 1921). The title sums up the content perfectly. Ernst Troeltsch's (1865–1923) *Der Historismus und seine Probleme* (*Historicism and its Problems,* 1922) attributed to Simmel the intention to overcome "suffocating 'historical realism.'" But Troeltsch was not satisfied with Simmel's reflections: "To Dilthey's relativism of values he [i.e., Simmel] adds a shaky relativism concerning the relationship between the reality of experience and the construction of historical knowledge. Historicism is overcome by means of its radical ... self-application; but really, it falls into a void, and nobody gains anything" (1922, 578, 582).

Not everyone agreed with Troeltsch. Using Simmel's historical *apriori* as a premise, Max Scheler (1874–1928) influentially declared historicism self-undermining (1926, 135–158). The historicist claimed the historical record to show that epistemic and moral standards vary between cultures and epochs. In so doing, the historicist "naïvely" assumed there to be "historical facts." This was naïve, since historicist reflection itself maintained that historical facts were "relative to the present." Historicism had thereby robbed itself of the data that could provide it with empirical support. Historicism refuted itself.

What did Simmel himself think of relativism and historicsm as philosophical positions? How did *he* interpret the historical *apriori*? Did he agree with Troeltsch that his position was a radical relativism?

To begin with the historical *apriori*, Simmel did not hold that all perspectives are equally good. For instance, while he was adamant that some aspects of historical actors' experiences might never be understood, he did not draw the relativistic conclusion that therefore historical sources were infinitely pliable. He added that when historians investigated the experiences of past *groups*—rather than past *individuals*—their chances of getting at the original sentiments were high. This was because groups had "general, big and coarse interests" that did not much change over time. Simmel also left room for the historian "genius" who was able to grasp past experiences to a higher degree than anyone else. The genius did so on the basis of "innate memories of the species." Moreover, Simmel wrote that "we are able to reconstruct the psychological processes of others and with a strong feeling of being totally correct ..."; or that "there is enough evidence" for the "monistic belief" that thought is largely "uniform and simple." Simmel was somewhat less optimistic when it came to historical laws or philosophies of history. Here, there was little chance to overcome or adjudicate competing proposals. Simmel weakened the relativism of these claims, however, by declaring historical laws and philosophies of history to be of merely heuristic value: as guides for research, they did not postulate conflicting truths (1892, 326, 328, 331, 337, 421).

As far as relativism in the *Philosophie des Geldes* is at issue, we need to attend primarily to the "relativity" point: the modern understanding of money corresponds to a diverse set of philosophical "relativisms." To begin with epistemic justification, Simmel held that it is relative to epistemic principles. And since these principles need to be justified in turn, we end up in an infinite or circular chain. Simmel actually insisted that the chain is both infinite *and* circular at the same time. It is infinite insofar as we are unable to reach absolute principles as ultimate stopping points. It is circular insofar as our justifying activity is never able to step beyond the circle of beliefs. (Simmel took these distinctions from Spencer [1867]). Somewhat abruptly, Simmel added an evolutionary epistemology according to which a representation is justified if, and only if, it has been produced by a psychological mechanism that adds to the fitness of the species. None of the "worldviews" of different species "copies" the external world "in its objectivity." They all have their own truth (1900, 101–104).

Simmel's ontological relativism related to the distinction between Medieval and Modern ways of thinking: the ontology of substances had given way to an ontology of (quantitative) relations. There was no fact of the matter as to which ontology is absolutely correct. Still, the thought of "the general relativity of the world" was the natural "adjustment on the part of our intellect" to contemporary "social and subjective life in which money has found its real effective embodiment and the reflected symbol of its forms and movements" (1900, 518). This line of reasoning made Simmel a historicist relativist about the conflict between (ontological) absolutism and relativism.

Finally, there was philosophical "relativism" concerning opposites like "pluralism" and "monism," "realism" and "idealism," "subjectivism" and "objectivism": Simmel treated them all as "heuristic principles." "Objective truth" resulted from the interplay of many different such principles (1900, 112).

Was all of this relativism by the standards of the time? Here, one key litmus test was—as the Neokantians or Phenomenologists were urging—whether the given theory conflated *Genese und Geltung* (causal origin and validity). Simmel declared himself "not guilty." He emphasized that the *law of gravity* "belongs within the category of the valid and meaningful that is not open to further [psychological or sociological] analysis" (1892, 105). Elsewhere, he sharply distinguished between "positive ethics" (i.e., the sociology and psychology of ethics) and "prescriptive ethics" (1892–93, 10–11). Max Adler (1919, 10) reported that "usually by relativism one means a state of mind which denies the possibility of universally valid claims and which reduces the validity of our judgments ultimately to their practical usefulness ..." Adler insisted that "nothing of all this fits Simmel's intellectual dispositions." Max Frischeisen-Köhler (1920) also suggested that "relativism" cannot be used to characterize Simmel's position. His reason was that relativism had become "too general and vague" a term for it to be useful as a characterization of any philosopher. Still, judged by Edmund Husserl's (1859–1938) attack on "psychologism," every form of evolutionary epistemology is "psychologistic" (i.e., species relativism) (1900, §36). Thus, by Husserl's criteria Simmel's evolutionary epistemology was relativistic.

Are Simmel's relativisms relativistic by our standards today? It seems right to say that, by most standards, Simmel would at least qualify as an ontological relativist. Which ontological theory we accept depends on the culture we live in, and there is no higher court of appeals. As far as epistemic relativism is concerned, the issue is difficult to decide. Simmel's insistence on infinite chains of justification, or his evolutionary speculations would not be counted as obviously relativistic by most of today's epistemologists. His references to the circle of beliefs sound like a form of coherence theory. It is an open debate whether and to what extent the coherence theory of epistemic justification has relativistic leanings (Steizinger 2015).

Summa summarum: It is hard to capture Simmel's enthusiastic search for ever new forms of relativism in today's categories and classifications. While one can recognize certain familiar relativistic motifs, the overall position remains elusive.

Mannheim on conservatism

Turning from Simmel to Mannheim, the first central text for my concerns is his *Habilitationschrift Altkonservatismus* (*Old Forms of Conservatism,* 1925). This is a study of four legal philosophers and jurists in the early eighteenth century: Justus Möser (1720–1794), Adam Heinrich Müller (1779–1829), Gustav von Hugo (1764–1844), and Friedrich Carl von Savigny (1779–1861). As Mannheim explained, they formulated conservatism as a new "thought-style," opposed to the "bourgeois-revolutionary style, the natural-law mode of thinking." A thought-style was motivated by social-political interests and "constituted a world" by means of its own specific vocabulary. Conservatism could, but did not need to, combine with romanticism. Romanticism arose in opposition to a "thorough-going rationalization of the world." It was motived by the "displaced" and "irrational." Political conservatism was an "objective," "historical-dynamic structural complex." It presupposed a "society differentiated into classes" (1925, 51, 56, 59, 65–66, 75, 86, 102).

One of Mannheim's central aims was to develop a "morphology of conservative thought," and to identify its "inner formative principle." By Mannheim's reckoning this principle was the *clinging to the concrete*. For instance, property (in Möser) was a "definite, vital, and reciprocal relationship between the owner and the thing owned." Freedom had to be concrete not abstract; it had to be tied to our "individual laws." Ranke later insisted that freedom must be tied to estates or the state (1925, 87, 89, 93).

Mannheim held that the conservative thought-style could only be understood as the negation of the earlier "natural-law thought-style." The latter was understood by conservatives as having the following ingredients (1925, 106–107):

(A) Doctrines of ...	(B) Thinking
State of nature	Rationalism
Social contract	Education from principles
Popular sovereignty	Universal principles
Rights of Man	Universal applicability of all laws
	Atomism and mechanism
	Static thinking

The conservative response took the form of insisting that social organisms are unique; this precluded universal principles or universal laws. It also favored totalities (cultures, traditions) over individuals. Against "reason," it stressed the importance of "history, life, nation," and celebrated the "irrationality of reality." To be precise, it was only the first generation of conservatives that attacked reason outright. Later generations—especially Hegel—aimed to fuse history and rationality by developing dynamic conceptions of reason (1925, 107–109).

Abstract morphology to one side, Mannheim also sought to illuminate the actual historical development. The "first conservative position" (in Möser) was a reaction of the Prussian nobility and its middle-class "ideologues" against "bureaucratic-absolutist rationalism." The ideologues were "socially unattached intellectuals" who "hired out their pens;" they were "arche-typal apologists." Importantly, (philosophy of) history became a key focus. For Mannheim, this was "the positive element" in conservatism. Möser emphasized old customs and habits; every town should have its own laws. Historicism emerged once the interest in history became pronounced. "Historicism is ... of conservative origin," Mannheim wrote. He explained that we reach historicism "when the process whereby things have come about is itself experienced with feeling." In this context, Mannheim applauded "... the fruitful relativism flowing out of historicism ... which renders even the observer relative to the process of becoming which moves over and through him." Mannheim added: "This form of thinking ... has in effect become a his-torical a priori for us" (1925, 112–114, 116–9, 126–127, 134, 143).

Müller was the central figure in the second stage of the development. He began to combine conservatism with romanticism, an emphasis on life and its diversity. For Mannheim, this move anticipated *Lebensphilosophie*. Müller was generally skeptical about conceptual thought. No concept was able to capture the dynamic nature of social realities. Müller aimed to make thinking similar to life: this meant an extensive use of analogy and conceiving of the world as unfolding according to diametrically opposed principles (1925, 136–138, 141, 143).

The third phase brought a consolidation of the conservative thought-style. This was the time of the "Historical School" and von Savigny. The latter never tired of attacking the idea of general (Napoleonic) legal codes: "The real seat of law is the common consciousness of the people." As a romantic conservative von Savigny discovered irrationality everywhere, including in the individual and in the application of laws. Mannheim deemed von Savigny important as a forerunner of the humanities of his (i.e., Mannheim's) time: von Savigny developed the method of "elucidation" (studying phenomena in and through their historical becoming) that informed all later work in the humanities (1925, 156, 166–167, 184).

A second important figure in the third stage of the development was von Hugo. He took "… a preliminary step towards the fruitful relativism of historicism" by developing a "unique relativism" tied to "disillusioned conservatism." von Hugo undermined natural law by measuring it against the plurality of positive law; but he also criticized positive law in light of natural law. Neither perspective was ultimately privileged. Von Hugo struck Mannheim as a forerunner of Max Weber (1925, 176).

Mannheim on historicism and relativism

Historicism and relativism surfaced at many points in the *Habilitationsschrift*, and it is useful to pull together the main themes and influences. First, historicism, relativism, and elucidation were said to be "positive elements," a valuable legacy, of political conservatism; a "historical a priori for us." Second, from Oswald Spengler's (1880–1936) *Untergang des Abendlandes* (*Decline of the West*, 1918), Mannheim adopted the idea of a "morphology" of "thought-styles," and of "formative principles" that "constitute" their respective "worlds" and incommensurable languages. Third, *Altkonservatismus* endorsed Simmel's conception of the historical *apriori*: Enlightenment and conservative writing of history were said to have been different; and at least by implication historical worlds had to be worlds-for-different-thought-styles. Finally, fourth, Mannheim did not allow for a neutral viewpoint. The fact that intellectuals were "socially unattached" did not enable them to achieve such neutrality; instead they were forced to "hire out their pens" (1925, 155).

In the paper *Historismus* (1924)—written just one year before the *Habilitationsschrift* was presented in Heidelberg—Mannheim drew a somewhat different picture of historicism and relativism. One central motif was the distinction between three different fields of knowledge and their respective histories (1924, 294). These fields differed in the extent to which they are affected by historicism.

(i) "Civilization": this was the realm of the natural sciences and mathematics, of "static truths" and progress (1924, 282). In this arena, and only here, the rationalist epistemologies of Neokantianism and Phenomenology had traction. These sciences were not affected by historicism.

(ii) "Dialectic Rationality": This was the domain of philosophy to which historicism applied. Here, truth was "dynamic," and could only be captured by Hegelian dialectic. At best, philosophical systems "express the truth of their respective epochs." But Mannheim also wrote of "dialectical truth" as the truth concerning the dynamics and telos of history (1924, 287, 289).

(iii) "Soul-culture": This was where the historical-cultural sciences belong. They were committed to the idea that "every epoch must be interpreted through its very own soul-center." This was in line with historicism or relativism. But still, Mannheim deemed it possible to offer a criterion of truth for this arena, too: it was to grasp the object "adequately in its full depth" (1924, 292–293).

A second central motif of the 1924 paper related to the "historical *apriori.*" Why do we assume that our perspectives are able to adequately capture the "historical *Dinge-an-sich?*" Mannheim thought that "Troeltsch found the right starting point" for an answer. Mannheim was referring here to Troeltsch's cryptic remark that "thought must ... have a secret link to reality": historians' perspectives were often adequate to their subject matter since both the perspectives and the subject matter were products of one and the same historical process. Mannheim expressed the thought also in idealist garb: "... principally the absolute ... can be grasped only ... in categories that are shaped by [its own] process of becoming"; "history gives us concrete-contentful standards; ... we are able to identify them only because we already have them instinctively, insofar as we are carried by the total spirit ..." (1924, 183, 276, 303–305)

For Mannheim the two central motifs—the distinction between (i), (ii), (iii) and the development of the absolute spirit—were connected. The absolute spirit had been invested in the three forms of knowledge to different degrees in different epochs. Mannheim also suggested that different social classes were preferentially drawn to one of the realms. But no class carried the "total movement" (1924, 296). Although Mannheim admitted that his own philosophy was also "tied to a standpoint," he still deemed himself capable of predicting the next step, beyond (iii). This next step was the challenge to find a perspective from which the tripartite distinction of fields of knowledge (i) to (iii) could be overcome.

This "pluralist" theoretical edifice, Mannheim believed, could be used to block the charge of relativism. Different perspectives could reach partial truths—insofar as the absolute granted these. Different truths of different historical periods were not different interpretations of the same, but interpretations of different stages of the absolute or spirit. This was not to say that perspectives could not be criticized. But the main criticism that could be directed at them was that they overgeneralized: they claimed to apply to a much wider range of phenomena than they actually did (1924, 253, 304).

In Mannheim's *Ideologie und Utopie* (*Ideology and Utopia*, 1931) we get yet a third response to historicism and relativism. Mannheim now sought to adapt Marxist vocabularies and motifs, especially elements of Georg Lukacs's *Geschichte and Klassenbewusstsein* (*History and Class Consciousness*, 1923). "Thought-styles" were replaced with "ideologies" *qua* holistic systems of meaning and beliefs. Mannheim now also put a much greater emphasis on the *Seinsgebundenheit* of thought (being bound to being). That is, the knowledge of politics, the humanities and the social sciences were all tied to material conditions. But Mannheim was convinced that the thesis of the *Seinsgebundenheit* of knowledge did not commit him to relativism. Mannheim preferred to call his view "relationism": "all historical knowledge is relational knowledge." In other words, different bodies of knowledge were always tied to different webs of meaning and beliefs, and to specific historical situations. This was why a strict division between *Genese* and *Geltung* could not be upheld. Of course, by the Neokantian standards, denying such strict division was precisely what qualified Mannheim's position as relativist (1931, 74, 76, 243–244).

Nevertheless, there was a criterion for assessing ideologies or theories, Mannheim claimed. A given ideology or theory was correct if it did not "prevent man from adjusting himself at the given historical stage." Finally, Mannheim needed an epistemic subject for identifying the specific *Seinsgebundenheit* of specific ideologies and for creating "syntheses" out of them. He resurrected a key figure from his *Altkonservatismus* study (and Weber [1923]): the "socially unattached [freischwebende] intelligentsia." Its epistemic privilege rested on its ability to replace class-ties with "Bildung" ties (1931, 85, 138).

Turning from Mannheim's views to their reception, in the late 1920s and early 1930s Mannheim's views triggered what in retrospect we might call the "sociologism wars." Since I have analyzed these debates at greater length elsewhere (Kusch 1999), and since I return to them below, suffice it to say that Mannheim's position was attacked from both ends of the political spectrum. Key points of contention were the restriction of *Seinsgebundenheit* to the social sciences and the humanities (Sombart); an excessive or insufficient proximity to Marx (A. Weber, Neurath); the inflationary use of "ideology" (Jonas, Stern, Tillich); the Hegelianism of the *Historismus* paper (Horkheimer); and—of course—Mannheim's alleged relativism (Grünwald).

These matters naturally bring us to the question of whether Mannheim was a relativist by today's standards. This is best answered separately for the three key texts. It is hard to see how one could read the *Altkonservatismus* study as anything but a relativistic investigation. The extensive use of Spenglerian categories and ideas committed Mannheim to a form of cultural relativism based on incommensurable thought-styles. The account of the *Historismus* paper avoided relativism, but at the high price of a Hegelian metaphysics. Finally, the position of *Ideologie und Utopie* escaped the relativism-charge only by relying on the assumption of the "free-floating intelligentsia."

Interpretation and evaluations

I now turn to some evaluative comments, using existing criticism as my starting points. I begin with problems in Simmel's case study of money and philosophy.

The first issue to be assessed is Simmel's relativistic methodology. Contemporary critics found it unconvincing. For Émile Durkheim (1858–1917), Simmel was wrong to isolate various forms and conceptions of money from the institutions that made them possible. Durkheim also condemned Simmel's "bastard speculation" as a "mixture of scientific observation and artistic intuition" (1902, 98). Max Weber (1864–1920), both in his *Die protestantische Ethik und der Geist des Kapitalismus* (*The Protestant Ethics and the Spirit of Capitalism*, 1904–1905) and in a 1908 manuscript (1970), was also critical. In the book, he rejected the way in which Simmel detached modern money from the capitalist mode of production (1904–1905, 193, 185). In the manuscript, Weber reported that the money-book triggered "outright explosions of rage" amongst economists. Weber saw the main culprit in Simmel's undisciplined use of analogy (1970, 160–161). It is hard to disagree.

Perhaps there is a way, however, to redeem Simmel's *Philosophie des Geldes*. Remember that Simmel took the book to be a contribution to the *philosophy of history* and that for him philosophies of history were not strictly true or false. Their value was to be measured by how useful they were as heuristic tools, that is, as temporary signposts for empirical work, and as ultimately and ideally fully superseded by the latter. Taken in that sense, we might think of Simmel as urging sociologists to determine the ways in which economic theorizing and financial practices have influenced philosophical thought. Treating Simmel's book in this way is to treat it the way we nowadays think of other such sweeping theses, for instance, the "Sombart thesis" (Sombart 1902) according to which double book-keeping brought about the Scientific Revolution; the "Weber thesis" (Weber 1904–1905) on the interaction between Protestantism and capitalism; or the "Merton thesis" (Merton 1938) on Pietism and early experimental science. These sweeping theses are overgeneralizations and false as such. But they are highly suggestive of more restricted and local case studies (e.g., on the influence of game-theory in political philosophy).

As far as Simmel's more general comments on relativism and historicism are concerned, the early responses were either highly critical or excessively charitable. I have already quoted Troeltsch's dismissive assessment above (1922, 582). More sympathetic commentators, like Adler (1919) or Frischeisen-Köhler (1920), confirm that Troeltsch's assessment was typical of the times. Curiously, Adler and Frischeisen-Köhler then went on to defend Simmel against the charge of relativism by rendering the doctrine in ways Simmel himself would have rejected: Simmel did not (like Frischeisen-Köhler) take relativism to be undefinable; and he did not (like Adler) conceive of relativism as a reduction of truth to utility.

The debate over Simmel's relativism did not end in the 1920s. Raymond Boudon suggests that Simmel was an anti-skeptical "neokantian relativist" seeking to identify "variable *apriori* assumptions" (1989, 415). Boudon's interpretation belongs to a French tradition of reading Simmel that Gregor Fitzi traces back to the early twentieth century. On this interpretation Simmel's relativism was a "transcendental philosophy aware of its limits in both epistemology and ethics" (2002, 247). One need not disagree with this suggestion to feel that it gets at only some aspects of Simmel's overall relativist aspirations. For instance, it has nothing to say about Simmel's ontological relativism.

Commentators from the realm of cultural studies highlight other aspects of Simmel's position. Deena and Michael A. Weinstein celebrate Simmel's "historicism" or "postmodern form of historicizing" as "a kind of radical pragmatism ... History is the freedom of the historian to historicize according to *any* interest in the past ... so long as the control of fact is respected" (1993, 183). Elizabeth S. Goodstein (2017, 86, 160) holds that "Simmel deployed epistemic relativism ... as a means of overcoming what appear to be aporetic dualisms—between materialism and idealism, determination and freedom, life and form ..." Goldstein believes that even the opposition between relativism and absolutism is overcome where Simmel's "relativized relativism" historicizes both. Goodstein's remark is indeed a helpful observation.

Finally, Klaus Christian Köhnke (1995, 480) and Wilfried Geßner (2003, 92–93) suggest that Simmel's relativism is really a form of Mannheimian "relationism." The reason, according to Geßner, is that Simmel's intentions are anti-skeptical and aiming to "combine the objectivity of validity with the relativity of origin." Unfortunately, neither author then goes on to defend the coherence and plausibility of Mannheim's relationism.

As already indicated in the last section, Mannheim's attempt to steer a path between relativism and absolutism did not meet with much enthusiasm in the 1920s. Max Horkheimer insisted that Mannheim should have followed Marx on this point. Whereas Marx had taken the view that all knowledge was relative to class, Mannheim was trying to re-introduce an absolutistic perspective. For Mannheim, all particular viewpoints ultimately seemed to fit together into one big absolute viewpoint (Horkheimer 1929, 486).

Ernst Grünwald chose a different line of attack. Grünwald accused Mannheim's sociology of knowledge of being both "absurd" and "false." The charge of absurdity was justified as follows: "'Relationism' claims that all thinking is valid only relative to a standpoint; but for this very sentence, that is, that all thinking is valid only relatively, relationism demands absolute validity." But Mannheim's "sociologism" was also based on the altogether false assumption that thinking in general, and judgments in particular were *seinsgebunden*. According to Grünwald, this had to be false: since judgments could be studied from many different viewpoints—e.g., those of sociology or psychology—judgments in themselves could not be reduced to any of these viewpoints. Finally, Grünwald argued that the sociology of knowledge could never replace epistemology and that research in the sociology of knowledge would always have to presuppose prior epistemological research. Epistemology studied the validity claims of judgments, and these validity claims were independent of, and prior to, external factors like social interests (1934, 701–707, 750). Mannheim did not have the argumentative resources needed to rebut these attacks.

There are also important and weighty objections to Mannheim's *Altkonservatismus* study. It is easy, for example, to agree with Rodney Nelson (1992) that there is "a relative paucity of social-historical analysis in Conservatism; [and that] a demographic portrait of the German aristocracy in the early nineteenth century and a rigorous documentation of shifts in their socioeconomic status would have added greatly to the analysis." Relatedly, Nelson also laments the "static essentialism of thought-styles" and the idea that "cultural products actually contain entelechies; that the development of a style of thought is in some indeterminable sense prefigured in its origin" (1992, 43, 45). Other recent critics attack the tautology in Mannheim's imputation of thought-styles to groups: Mannheim defined groups in terms of thought-styles, and thought-styles in terms of groups (Carlsnaess 1981). On a more general level, latter-day sociologists of knowledge lament Mannheim's excluding of mathematics and the natural sciences from the subject matter of their discipline (Bloor 1973; cf. Seidel 2011).

And yet, all these important criticisms do not invalidate the attempt to define political conservatism; and Mannheim's definition remains influential even today. Moreover, the essential core of Mannheim's sociohistorical study of philosophical knowledge can easily be formulated without any "morphology of thought-styles." As an exercise in the sociology of philosophy—showing how historicism and relativism first emerged in political conservatism—Mannheim's *Conservatism* remains a model and inspiration.

Turning to the broader issue of how to think about historicism and relativism, I want to underline for a last time the interesting historical reflexivity in both accounts. Both Simmel and Mannheim seek to understand the historical conditions of the possibility of their own analyses—and this for both historicism and relativism. This type of historical reflexivity is a rare commodity in contemporary sociology of philosophical (or scientific) knowledge.

Moreover, Simmel's contribution to historicism is still thought-provoking and well worth reflecting on. That historical facts are not simply given is now widely taken for granted; and today we have of course authors like Hayden White (1973) and Ian Hacking (1995) who put forward more radical historiographical claims. Of course, philosophers today do not believe that historical reflexivity invariably leads into an abyss—like Troeltsch did; nor that it can justify universal values—like Scheler did. But we still struggle to explain how historical reflexivity and historical realism are to be reconciled. In that sense, Simmel's challenge is still pressing today.

Finally, as far as Simmel's celebration of relativism is concerned, it is a bit unfortunate that he ran so many different ideas together under this one title. Many of the things Simmel called "relativism" are today flourishing research projects in their own right (e.g., evolutionary epistemology, pluralism, and coherence theories).

Turning from Simmel to Mannheim, the latter's contortions to escape the charge of relativism are hardly convincing, since in each of his programs the intellectual costs are too high. Spenglerism, with its ingredients of incommensurability, skepticism, and organicism, is not an attractive option. The Hegelianism of the *Historismus* paper avoided epistemic relativism only at the cost of ontological relativism. (The absolute presents different periods with different realities.) Concerning the free-floating intelligentsia, it is puzzling how Mannheim by 1931 could suddenly convince himself that it is able, after all, to escape "the fate" so vividly described in *Altkonservatismus*.

Summary and conclusions

I have tried to show Simmel's and Mannheim's deep and extensive preoccupations with relativistic and historicist themes. Both philosopher-sociologists thought of relativism and historicism as important elements of their methodology for the sociology of philosophy; both tried to understand the emergence of relativism and historicism historically; and both reflected systematically on which forms of relativism might be defensible. Many of their central questions and concerns have stood the test of time.

There are also, of course, important differences between Simmel and Mannheim. On the one hand, as we move historically forward, from Simmel to Mannheim, the sociology of knowledge sheds a commitment to broad-brush and speculative philosophy of history, at least as far as the case studies are concerned. This is progress. On the other hand, the early Simmel's theorizing concerning historicism seems more sober, more twenty-first-century than Mannheim's frequent relapses into Spenglerism or Hegelianism (at least in his programmatic texts). At the same time, in his theoretical writings Mannheim challenges us to investigate how epistemology must change to do justice to the results of the sociology of knowledge. To me, the question is still relevant.

Where does all of this leave us concerning the issue of relativism, especially relativism in the context of the sociology of knowledge? The question is worth asking in this general form, since clearly the issue is as topical today as it was 100 years ago (cf. Schantz and Seidel 2011). Obviously, we cannot simply adopt "the right answer" from Simmel and Mannheim; as we have seen, their proposals are beset by many problems. But perhaps the real lesson is, in any case, an indirect one: perhaps the connection between a philosophically worked-out commitment to relativism on the one hand, and historical case studies in the sociology of knowledge on the other hand is not as tight as is often assumed by card-carrying sociological relativists. Simmel declared himself a relativist, but it remained unclear what precisely his relativism amounted to. And yet, he gave us a highly suggestive and thought-provoking investigation into the influence of (theorizing about) money on our modern worldview. Mannheim had a clearer view of what relativism was but thought of it as something the sociology of knowledge had to avoid at all costs. His attempts to do so failed. But this failure had no direct negative consequences for his brilliant historical work. Fortunately, in neither case did the lack of success regarding the philosophical handling of relativism weaken the quality and interest of the sociological-historical case studies.

Acknowledgments

Work on this chapter was supported by European Research Council Grant #339382, "The Emergence of Relativism." I am also indebted to Katherina Kinzel, Johannes Steizinger, and Niels Wildschut for many helpful comments and corrections. Last but not least, I am grateful to audiences in Geneva and Vienna.

References

Adler, M. (1919), *Georg Simmels Bedeutung für die Geistesgeschichte*, Leipzig: Anzengruber.

Bloor, D. (1973), "Wittgenstein and Mannheim on the Sociology of Mathematics," *Studies in History and Philosophy of Science* 4: 173–191.

Boudon, R. (1989), "Die Erkenntnistheorie in Simmels 'Philosophie des Geldes'," *Zeitschrift für Soziologie* 18: 413–425.

Carlsnaes, W. (1981), *The Concept of Ideology and Political Analysis*, Westport, CT: Greenwood Press.

Fitzi, G. (2002), *Soziale Erfahrung und Lebensphilosophie: Georg Simmels Beziehung zu Henri Bergson*, Konstanz: UVK Verlagsgesellschaft.

Frischeisen-Köhler, M. (1920), "Georg-Simmel," *Kantstudien* 24: 1–51.

Geßner, W. (2003), *Der Schatz im Acker: Georg Simmels Philosophie der Kultur*, Weilerswist: Velbrück.

Goodstein, E. S. (2017), *Georg Simmel and the Disciplinary Imaginary*, Stanford, CA:. Stanford University Press.

Grünwald, E. (1934), "Systematische Analyse der wissenssoziologischen Theorien," in Meja and Stehr (1982), 681–747.

Hacking, I. (1995), *Rewriting the Soul: Multiple Personality Disorder and the Sciences of Memory*, Princeton, N.J.: Princeton University Press.

Horkheimer, M. (1930), "Ein neuer Ideologiebegriff," in Meja and Stehr (1982), 474–496.

Husserl, E. (1900), *Logical Investigations*, translated by J. N. Findlay, London: Routledge & Kegan Paul, 1970.

Kusch, M. (1999), "Philosophy and the Sociology of Knowledge," *Studies in History and Philosophy of Science* 30A: 651–586.

Köhnke, K. C. (1995), *Der junge Simmel in Theoriebeziehungen und sozialen Bewegungen*, Frankfurt am Main: Suhrkamp.

Lessing, T. (1921), *Geschichte als Sinngebung des Sinnlosen*, Bremen: DOGMA, 2012.

Lukacs, G. (1923), *History and Class Consciousness: Studies in Marxist Dialectics*, translated by R. Livingstone, Cambridge, MA: The MIT Press, 2000.

Mannheim, K. (1924), "Historismus," in Mannheim, *Wissenssoziologie: Auswahl aus dem Werk*, introduced and edited by K. H. Wolff, Berlin: Luchterhand, 1964, 246–307.

—— (1925), *Conservatism: A Contribution to the Sociology of Knowledge*, edited by D. Kettler, V. Meja and N. Stehr, translated by D. Kettler, V. Meja, and E. R. King, London: Routledge, 1986.

—— (1931), *Ideology and Utopia*, translated by L. Wirth and E. Shils, London: Routledge & Kegan Paul, 1936.

Merton, R. K. (1938), "Science, Technology and Society in Seventeenth-Century England," *Osiris* 4: 360–632.

Meja, V. and N. Stehr (eds.) (1982), *Der Streit um die Wissenssoziologie*, 2 vols., Frankfurt am Main: Suhrkamp.

Nelson, R. D. (1992), "The Sociology of Styles of Thought," *The British Journal of Sociology* 43: 25–54.

Schantz, R. and M. Seidel (eds.) (2011), *The Problem of Relativism in the Sociology of (Scientific) Knowledge*, Frankfurt am Main: Suhrkamp.

Scheler, M. (1926), *Die Wissensformen und die Gesellschaft, Gesammelte Werke,* vol. 8, edited by Maria Scheler, Bern und München: Francke, 1960.

Seidel, M. (2011), "Karl Mannheim, Relativism and Knowledge in the Natural Sciences—A Deviant Interpretation," in Schantz and Seidel (2011), 183–214.

Simmel, G. (1892), "Die Probleme der Geschichtsphilosophie," in *Gesamtausgabe*, vol. 2, edited by H.-J. Dahme, Frankfurt am Main: Suhrkamp, 1989, 297–421.

—— (1892–1893), *Einleitung in die Moralwissenschaft,* in *Gesamtausgabe*, vols. 3 and 4, edited by Klaus Christian Köhnke, Frankfurt am Main: Suhrkamp, 1989 and 1991.

—— (1900), *The Philosophy of Money*, edited by D. Frisby, translated by T. Bottomore and D. Frisby, London: Routledge, 2004.

—— (2005), *Briefe 1912–1918: Jugendbriefe,* in *Gesamtausgabe,* vol. 23, edited by O. and A. Rammstedt, Frankfurt am Main: Suhrkamp.

Sombart, W. (1902), *Der moderne Kapitalismus*, Berlin: Duncker & Humblot.

Spengler, O. (1918), *The Decline of the West*, translated by C. F. Atkinson, New York: Oxford University Press, 1991.

Spencer, H. (1867), *First Principles*, London: Williams and Norgate.

Steizinger, J. (2015), "In Defence of Epistemic Relativism: The Concept of Truth in Georg Simmel's Philosophy of Money," in *Realism—Relativism—Constructivism*, edited by C. Kanzian, J. Mitterer and K. Neges, Kirchberg am Wechsel: Austrian Ludwig Wittgenstein Society, 300–302.

Troeltsch, E. (1922), *Der Historismus und seine Probleme*, reprint, Aalen: Scientia Verlag, 1977.

Weber, A. (1923), *Die Not der geistigen Arbeiter*, München, Leipzig: Duncker & Humblot.

Weber, M. (1904–1905), *The Protestant Ethic and the Spirit of Capitalism*, translated by T. Parsons, London and Boston: Unwin Hyman, 1930.

—— (1970), "Georg Simmel as Sociologist," *Social Research* 39: 155–163.

Weinstein, D. and M. A. Weinstein (1993), *Postmodern(ized) Simmel*, London and New York: Routledge.

11 Was Heidegger a relativist?

Sacha Golob

Introduction: The systematic importance of Heidegger on relativism

At first glance, it might seem that relativism was not a central issue for Heidegger. He was, of course, extremely familiar with the Husserlian and post-Kantian debates that linked relativism, logic, and psychologism: these had been the focus of his 1914 dissertation, *Die Lehre vom Urteil im Psychologismus* (*The Doctrine of Judgment in Psychologism*). There, he attacks psychologism in part by linking it to relativism, as Husserl had done before him: The challenge, Heidegger suggests, is to move beyond this negative point and to articulate a positive story about judgmental content, one that respects the phenomenology of the act of judging in a way that Husserl's own theory allegedly does not (Ga 21, 107, 111).[1] But while Heidegger presses the issues of judgment and content closely in his mature work, by the time we reach *Sein und Zeit*, there is little explicit treatment of relativism. At points, Heidegger uses it there simply as a byword for philosophical error: he is quick to insist that his views have nothing to do with a "*schlechte Relativierung*" (crude relativizing). He warns against readings of Dilthey as offering a "relativistic" *Lebensphilosophie* (philosophy of life), and he praises Yorck for seeing through "all 'groundless' relativisms" (SZ 22, 399, 401). Admittedly, he does state that "all truth is relative to *Dasein*'s being," but, having clarified that this does not mean that truth is "left to the subject's discretion," he promptly drops the term and does not take it up again (SZ 227). Similarly, in other works, both before and after SZ, relativism is directly treated only in marginal contexts. The 1921 lectures *Phänomenologische Interpretationen zu Aristoteles* (*Phenomenological Interpretations of Aristotle*), for example, warn against the "atrophy of relativism" when discussing the link between philosophy and the university (Ga 61, 69), while 1935's *Die Frage nach dem Ding* (*The Question Concerning the Thing*), perhaps the clearest treatment of the Galilean paradigm shift, uses *Relativismus* only once, in a dismissive survey of standard views on indexicals. As elsewhere, it is clear that Heidegger regards both the term, this "cheap label," and the typical reactions to it, as problematic (Ga 41, 28).

Appearances can be misleading, however. Relativism is, in fact, fundamental to understanding Heidegger's philosophy and its place in the canon. There are four reasons for this.

First, Heidegger's refusal to rehash the standard debates around relativism is motivated by the belief that these are symptoms of a series of underlying errors (see, for example, Ga 21, 21–22 on the textbook self-reference arguments). By mapping his stance on relativism, we can get a better feeling for what those errors were, and how he sought to move beyond them. Second, while Heidegger avoids the usual terminology and framing, many of the issues raised by relativism reoccur in his work on truth and in his epistemology, and need to be addressed if that work is to be articulated and defended. Third, Heidegger's relation to relativism is particularly important for understanding how his philosophy relates to some of the central tensions in the post-Kantian tradition. For example, to what degree can one really combine the transcendental language that SZ borrows from the first *Critique* with Heidegger's post-Hegelian emphasis on history and his existentialist interest in facticity? Does the existential analytic yield anything like a universal transcendental framework or is there at best a series of "historical *a prioris*"? How stable is the divide between ontological and ontic knowledge? All of these questions are illuminated by approaching them from the relativism angle. Fourth, the question of relativism decisively colors the early reception of Heidegger's work. Husserl's 1931 lecture *Phänomenologie und Anthropologie* warns explicitly against a philosophy based on the "essence of human being's concrete worldly *Dasein*": this approach can only lead to "anthropologism," the pejorative term used for species-relativism in the *Prolegomena zur reinen Logik* (*Prolegomena to Pure Logic*) (Husserl 1997, 485). Husserl's charge, roughly, is that a philosophy founded on a study of human beings will relativize logic to facts about such beings and their mental capacities. While Heidegger is not mentioned by name, it is clear that he is the target here: Husserl positions him as combining a rhetoric of authenticity and historicity with the kind of psychologism the *Prolegomena* had attacked in Benno Erdmann thirty years earlier (Husserl 1975, §§38–41). Such "*Dasein* anthropology" "constitutes a complete reversal of phenomenology's fundamental standpoint" (Husserl 1997, 486). To assess the accuracy of this charge, and Heidegger's place in the phenomenological tradition, we need to know where he stands on relativism.

The structure of this chapter will be very simple. In the first half, I will introduce a sophisticated way of reading Heidegger as a relativist; I draw here on the work of Kusch and Lafont. In the second half, I present the counter-argument. As I see it, Heidegger is not a relativist; but understanding the relations between his approach and a relativistic one is crucial for an evaluation of both his own work and the broader trajectory of post-Kantian thought.

Before proceeding, a brief caveat: Heidegger was a prolific writer: the *Gesamtausgabe* edition runs to over 100 volumes. Furthermore, during the course of his lifetime, his work undergoes a series of complex stylistic and philosophical shifts—for example, during the early 1930s and then again in the aftermath of the war. There is no scholarly consensus on the exact nature of these developments or on the degree of continuity or change that they imply. Given these facts, it would be impossible to address Heidegger's views on "relativism" or indeed any other topic in a single article without radically restricting the chronological range of the discussion. I will therefore focus on Heidegger's best-known work, *Sein und Zeit* (*Being and Time*) (1927), and on the account developed there and refined in subsequent texts. In this sense, what follows is largely, although by no means exclusively, a study of "early Heidegger"; for stylistic reasons, I will refer simply to "Heidegger," taking the qualification as understood.

Heidegger as relativist: Lafont's reading

Explicitly relativist readings of Heidegger have been advanced by both Kusch and Lafont (Kusch 1989; Lafont 2000, 2007). In what follows, I will focus on Lafont's account, both because of its large influence on the recent secondary literature and because it explicitly brings out the connections to Kant that I think are crucial.[2]

Lafont's basic claim is that Heidegger is a "conceptual scheme" relativist who holds that "truth is relative to a prior understanding of being" (Lafont 2002, 187). There are, in fact, two important issues here. The first is the assumption that one can equate "world-disclosures, understandings of being, conceptual schemes" (Lafont 2002, 187). For Lafont, Heideggerian understanding is tacitly propositional and closely related to the predicative structure found in language (Lafont 2000, 181 n1); Kusch likewise describes Heidegger as a "linguistic relativist" (Kusch 1989, 21, see also 196–197). In this, Lafont and Kusch differ markedly from the standard view on which Heidegger's achievement was precisely to break with what Carman called the "assertoric paradigm," the tacit modelling of meaning on language (Carman 2003, 216). This debate, while vital for a broader understanding of Heidegger, would take us too far afield here, and so I set it aside.[3] Instead, I want to focus on the second aspect of Lafont's approach, the relativization claim. To understand that, we need to begin by looking at Lafont's treatment of the *a priori*.

On Lafont's model, Heidegger's relativism is primarily a function of his attempt to combine a quasi-Kantian story about the *a priori* with an increased emphasis on history. The basic idea is as follows. On Kant's picture, there is a single, universal *a priori* structure shared at least by all human agents.[4] For Heidegger, in contrast, there are supposedly multiple *a priori* structures, each tied to a particular historical period. As Lafont puts it:

> Understanding of being is not the (eternal) endowment of a transcendental ego ... but is merely contingent, changes historically and cannot be put under control at will. It is thus a fate into which human beings are thrown.
>
> (Lafont 2002, 186)

These structures—"world-disclosures, understandings of being, conceptual schemes"— determine how we experience entities, just as for Kant the *a priori* conditions of human understanding and sensibility condition appearances. By extension, all truths about the objects of experience are relative to those structures (Lafont 2007, 105).

Heidegger is thus a relativist in a double sense. First, "truth is relative to a prior understanding of being": this is his appropriation of transcendental idealism (Lafont 2007, 105). Second, his work is defined by the multiplication or "relativization of the Kantian conception of apriority," that is his willingness to historicize the *a priori*, recognizing different epochs, each with its own understandings of being (Lafont 2007, 118). The result is a relativism driven by a basically Kantian framework, within which Heidegger simply "substitutes the ontological difference for the empirical/transcendental distinction" (Lafont 2000, xii).

The question of how the *a priori* interacts with historical change is, of course, central to post-Kantian thought from Hegel onward: as Foucault observed, talk of a "historical *a priori*" produces "a rather startling effect" (Foucault 1971, 127). Heidegger was extremely familiar with such issues from his work on Dilthey, whose ambition was nothing less than a "critique of historical reason": indeed, Kisiel labels the 1924 draft of SZ the "Dilthey Draft," such is the extent of the influence (Dilthey 1988, 141; Kisiel 1993, 315). The idea of some kind of historical *a priori*, and the complexities that brings with it, is very clearly present in the key texts cited by Lafont. For example, when discussing the shift to modern mathematical physics, Heidegger describes the Galilean revolution in terms close to a Kuhnian "paradigm-shift": Galileo's achievement was not in any sense straightforwardly empirical; rather, he set up a new model or "projected plan" in terms of which entities could be then interpreted, calculated, and predicted (Ga 41, 89–91). This type of framework:

> Determines in advance the constitution of the being of entities ... This prior plan of the being of entities is inscribed within the basic concepts and principles of the science of nature.
>
> (Ga 3, 11)

Underlying this move is an equation of such "paradigms" with synthetic *a priori* judgments, themselves understood as transcendental principles in terms of which and through which we encounter entities (Ga 41, 183–184).[5]

What is distinctive about Lafont's Heidegger is the rigorous and systematic way in which she elaborates these initial moves. I want to highlight three dimensions of her approach in particular.

First, each instance of the *a priori* retains the full determinative force of its Kantian predecessor. So, for example, both the Aristotelian and Galilean ontologies retain "the absolute authority ... that *a priori* knowledge is supposed to have" (Lafont 2007, 107).

It is vital to see that for Lafont, this is not simply an epistemic claim: it is not the claim that people find it hard, or perhaps even impossible, to escape from the assumptions that define our period or its best science. Rather, it is a *constitutive* claim, one that commits Heidegger to what is effectively a historicized transcendental idealism in which one set of appearances, those of the Greek world, is suddenly replaced by another, those of the modern one.

> The way in which entities are understood must determine in advance which entities we are referring to or, in general terms, meaning must determine reference ... Given that the prior understanding of the being of entities is what makes our experience an experience of some specific entities (rather than others), it determines what these entities are (for us), that is, it determines what they are accessible to us as ... The understanding of the being of entities determines all experience of those entities.
>
> (Lafont 2007, 108, I have inverted the order of
> the final two sentences)

The ontological difference is thus:

> A dichotomy in which one pole (the meaning pre-given in an understanding of being) necessarily assumes constitutive powers over the other (i.e., over our access to the referents, to the intraworldy entities).
>
> (Lafont 2000, 180)

One sees here the links between Lafont's relativism and her understanding of Heidegger's philosophy of language: she thinks that the constitutive role of ontological knowledge is ultimately underwritten by his commitment to a theory of "indirect reference" (Lafont 2000, 181).

Second, it follows that there exists an extreme incommensurability among the various frameworks. Since each understanding of being is "responsible for the constitution of objects ... an alternative projection is (by definition) a projection of different objects and thus incommensurable with it" (Lafont 2000, 171). This, Lafont argues, is the point Heidegger is making in texts such as this:

> It is simply useless to measure the Aristotelian doctrine of motion against that of Galileo with respect to results, judging the former as backward and the latter as advanced. For in each case, nature means something completely different.
>
> (Ga 45, 52–53; cited by Lafont in her 2007, 111)

In other words, each of the various *a prioris* constitutes a genuinely distinct set of entities: claims about entities can only be true or false relative to that framework.

> Given that, according to Heidegger, entities are only accessible through a prior projection of their being, it is clear that entities made accessible by genuinely different projections are, by definition, not the same entities.
> (Lafont 2007, 112)

Kusch's Heidegger, although motivated more by reflections on language and less by direct links to Kant, similarly sees us as "trapped in our project" (Kusch 1989, 238).

Third, since the understanding of being determines the nature of the entities we can encounter, there is no possibility of that encounter forcing any revision in that understanding. For Lafont's Heidegger, there is therefore a substantive class of claims, including the basics of both Aristotelian and Galilean physics, which are immune to empirical correction. In line with the incommensurability just discussed, these same principles cannot be criticized from any external perspective, since the reference of any theory is determined entirely by the beliefs that constitute that theory: those who disagree are thus necessarily not talking about the same thing (Lafont 2007, 112, 117). This, unsurprisingly, has far-reaching epistemic implications. For example, Lafont sees Heidegger as unable to make sense of the standard idea of scientific progress, insofar as that entails the gradual revision of our principles based on their empirical testing: "Thus, the attempt to conceive the historical changes in our understanding of being as a learning process is based on an illusion ... They are unrevisable from within and inaccessible (meaningless) from without" (Lafont 2007, 112).

We now have a fairly detailed account of Lafont's views in place: what should we make of them? There are, as ever in the history of philosophy, two questions: is this an intellectually viable position, and was it Heidegger's position? What is striking is that both Lafont and her opponents agree on a negative answer to the first question. For Lafont, the significance of Heidegger is ultimately as a cautionary tale, warning against the "indirect theory of reference," which supposedly lay behind his approach; he should, instead, have opted for some kind of direct reference story, perhaps of a broadly Kripkean type (Lafont 2000, xvii, 184). Other commentators, who typically agree with Lafont on little else, concur in this philosophically negative verdict. Wrathall, for example, describes the theory's underlying assumptions regarding both reference and scientific progress "as patently absurd" (Wrathall 2011, 121). I do not, however, want to approach the issue in terms either of reference or scientific progress: Heidegger's views on both are changeable and hang on myriad subsidiary questions, such as whether philosophy is a science.

Instead, I want to make a more direct move. I think that the picture Lafont paints of Heidegger, while highly sophisticated, is also deeply mistaken. My aim in the second half of this chapter will be to advance an alternative reading, one that has at least as much textual support and avoids the philosophical dangers of Lafont's model. In line with the principle of charity, we should attribute this second view to Heidegger—as we will see, it is one on which he stands fundamentally opposed to relativism.

Heidegger as anti-relativist: The hermeneutic reading

I will now argue that Heidegger is not a relativist. Instead, his position is that relativism is an understandable, but ultimately misguided, response to the errors of its dialectical opponents. As he put it himself:

> The theories of relativism and scepticism originate in a partly justified opposition against a distorted absolutism and dogmatism with respect to the concept of truth.
>
> (Ga 24, 316)

Heidegger's own preferred tactic will be to put into question the underlying assumptions that have left us oscillating between the two poles of relativism and absolutism.

Before I can address Heidegger's position directly, however, I need to clear some other issues out of the way. The basic problem is that his views on relativism are interwoven with his stance on some of the most contested and complex topics in his thought: truth, idealism, and being. Clearly, I cannot treat all of these here; my aim instead is to get to a point where the key questions for current purposes can be isolated and focused on.

Truth

Heidegger clearly defends a view of truth that is relational in some important sense: as he puts it in a famous passage, in the absence of *Dasein*, Newton's laws would not be true (SZ 226–227). They would not be false either: rather, it is simply inappropriate to talk about truth or falsity under that condition (226–267). As Kusch perceptively notes, this type of relationality falls short of relativism; given his own relativistic reading of Heidegger, Kusch attributes this to a reluctance to state the doctrine openly:

> It seems that Heidegger should have gone further here by saying that Newton's laws are true only for those *Daseins* that share the same universal medium of meaning with Newton; yet the Heidegger of *Being and Time* does not seem to be ready to state the relativism of his notion of truth so bluntly.
>
> (Kusch 1989, 191)

As I see it, however, the key is that the dependence of truth on *Dasein* will not entail relativism in any significant sense if all relevant *properties* of entities remain independent of *Dasein*. Suppose, for example, that the properties and behaviors identified by Newton's laws were to remain exactly as they are, even if *Dasein* had never existed or if *Dasein* had always endorsed an Aristotelian view, and it is simply that one can only declare the laws "true" insofar as someone believes them. What we would then have is a proposal to modify the use of the truth predicate rather than any substantive relativism. Indeed, this is precisely what I have argued elsewhere: Heidegger's remarks on truth are primarily motivated by the phenomenological assumption that "being true is a comportmental relation between the presumed and intuited, namely identity" (Ga 20, 70).[6] We do not need to settle this here; what matters is simply that there is a non-relativistic way of reading Heidegger's remarks on Newton.

Idealism and being

The relativist might concede that this is correct, but argue that the problem is with the realist assumptions in the antecedent of my conditional: after all, for someone like Lafont, the properties of entities do *indeed* change along with *Dasein*'s views. In this way, the focus shifts from truth to idealism. My own view is that Heidegger was a realist, and I think the exegetical situation is much more complex than Lafont recognizes, partly because Heidegger, perhaps uniquely, also viewed *Kant* as a realist (Golob 2013). But again, this is not the place to debate those issues: one fundamental issue, as Blattner has stressed, is that the ambiguity between realism and idealism is present in the very definition of "being" used by Heidegger.

On the one hand, as "being" is traditionally used to mean something like "that in virtue of which an entity is an entity and an entity of the sort it is" (Blattner 1999): thus, Heidegger himself introduces "being" as "that which determines entities as entities" (SZ 6). On the other, Heidegger immediately moves to a definition of "being" as "that in terms of which entities are already understood" (6). Carman thus identifies "being" as "the condition of the intelligibility of entities as entities", while Frede similarly glosses "to be" as "to be understood as" (Carman 2003, 15; Frede 1993, 57). The result is that to draw the balance between the realist and idealist strands of Heidegger's project one would need first to get clear on his conception of being, and indeed on his understanding of Kant. But again, all we need for current purposes, is to note that a realist reading is by no means impossible: there are, for example, plenty of passages that accord with a straightforward realism on which all the plausibly mind-independent properties of entities are indeed mind-independent. Consider these:

> World is only, if, and as long as *Dasein* exists. Nature can also be when no *Dasein* exists.
>
> (Ga 24, 241)
>
> Entities are in themselves the kind of entities they are, and in the way they are, even if, for example, *Dasein* does not exist.
>
> (Ga 26, 194)

As with truth, I cannot settle this debate here: my point is simply to show that they are indeed *debates*, and thus that there is space for something other than a relativistic reading. The task now is to return to relativism itself and to sketch out how such a reading might look.

Back to relativism

The place to start is with a basic point made by Wrathall against Lafont:

> If Heidegger were simply advancing the weaker hypothesis that whenever we experience anything, "we have always already understood entities in one way or other," his claim would be unobjectionable. But [Lafont] sees him as advancing the much stronger thesis that "the way in which we in fact have always already understood everything is constitutive of what things are or of what things we can refer to" (Lafont 2000, 139, n31) … No one would deny that Heidegger believes our experience of things is *guided* by a meaningfully structured understanding of the world.
>
> (Wrathall 2002, 219–220, original emphasis)

Heidegger certainly thinks that an understanding of being must precede any encounter with entities: indeed, this is how he defines the Copernican Turn (Ga 3, 13). However, as Wrathall observes, that need not imply that this understanding *constitutes* entities in the very strong sense that Lafont relies upon. Instead, it might simply shape our encounter in some much weaker sense.

How should we develop this basic point? Heidegger himself provides a detailed answer. Given the importance of the following passage, I quote in full:

> It was an error of phenomenology to believe that phenomena could be correctly seen merely through unprejudiced looking. But it is just as great an error to believe that, since perspectives are always necessary, the phenomena themselves can *never* be seen and that everything amounts to contingent, subjective anthropological standpoints. From these two impossibilities, we obtain the necessary insight that our central task and methodological problem is to arrive at the *right* perspective. We need to take a preliminary view of the phenomenon but precisely for this reason it is of decisive importance whether the guiding perspective is adequate to the phenomenon, i.e., whether it is derived from its substantial content or not (or only constructed). It is not because we must view it from some perspective or other that the phenomenon gets blocked off to us, but because the perspective adopted most often does not have a genuine origin in the phenomenon itself.
>
> (Ga 34, 286, original emphasis)

Heidegger here rejects the fantasy of a view from nowhere. We always approach entities in terms of some understanding of being; we always operate out of some specific hermeneutic situation. But those assumptions do *not* blankly determine the entity as on Lafont's picture. Rather, "our central task" is to engage in a continuous process of adjusting and recalibrating our standpoint in order "to arrive at the right perspective."[7] The "mode of discovery" must be "as it were, regulated and prescribed by the entity to be discovered and by its mode of being" (Ga 24, 99). Thus, the key task is to "secure the right access" to the entities we are interrogating (SZ 15), the right starting point for the inquiry, the right methods (SZ 36). All the while, we need to be conscious that many familiar principles or concepts or tools will be unsuitable because they are not sufficiently attentive to the dynamics of the domains in question (SZ 36); for example, one cannot simply appeal to modal logic without recognizing that the notions of modality appropriate to different entities are not even coextensive (SZ 143–144). In order to establish a "stable way of coining the appropriate concepts" (SZ 55), we therefore need, in classic hermeneutic fashion, to first become aware of the baggage, the imbalances and prejudices, which the tradition has bequeathed us (SZ 22). To do this, SZ seeks to identify certain systematic source of error, such as *das Man,* and certain systematic devices for escaping such error, such as anxiety. The result is not a relativistic picture but one that is both deeply phenomenological and deeply hermeneutic: the key "methodological problem" is precisely how to *develop* our understanding of being so that it allows the phenomenon to "show itself from itself."

One reason this result has been missed, I would suggest, is a tendency by commentators to conflate regional and fundamental ontology: the contrast is not mentioned, for example, in the Lafont article that sets out her view in most detail (Lafont 2007). As I see it, Heideggerian fundamental ontology, for example the claim that *Dasein* encounters entities by locating them within a teleologically structured world, has a classically *a priori*, universal, and transcendental status: it holds for all *Dasein.* By extension, there is no question of adjusting it to the entity in question: it is unchangeable. But this is unproblematic. Not only is it universal, thus preventing the proliferation of frameworks that so troubled Lafont, but it concerns only properties that are obviously relational: not even the most ardent realist would have a problem with the fact that something's being "equipment" depends on the existence of *Dasein.* But when it comes to regional ontologies such as Aristotelian or Galilean science, ontological truth is, to borrow a phrase from Haugeland, "beholden to entities" (Haugeland 2013, 201): it is, and must be, open to adjustment and revision as we seek to "arrive at the *right* perspective" (Ga 34, 286, original emphasis).

To develop this approach further, I want to address two issues in particular. The first concerns empirical correction in the natural sciences. As you will recall, on Lafont's account, Heidegger is committed to a near-complete rejection of empirical inquiry: all that such a process amounts to is the "playing out" of the "axioms" that pre-define that understanding of being (Lafont 2000, 286). On my reading, in contrast, things are much more fluid. Heidegger certainly doubts that first-order scientific study will be enough to shift dominant, but mistaken, paradigms. But this is not because our understandings of being are inherently "unrevisable from within and inaccessible (meaningless) from without" (Lafont 2007, 112). It is, rather, because a *mixture* of methods is needed to adjust our perspectives in the right ways. For example, Heidegger criticizes the natural sciences, in particular modern mathematical physics, by arguing that their key concepts are drawn from historically questionable sources (SZ 362, Ga 41, 33, 92–93). For Heidegger, these methodological failings imply that such sciences are, in an important sense, not genuinely attending to the phenomena themselves: for all their stress on experimentation, they are not truly engaging with the data, but rather are driven by antecedent assumptions to "skip over the facts" (Ga 41, 93).[8] This is how we should read passages such as the following:

> The Greek doctrine of natural processes does not rest upon insufficient observation, but rather upon a different (and perhaps even deeper) concept of nature that is prior to all particular observations.
>
> (Ga 45, 52)

This passage was cited by Lafont in favor of her approach (Lafont 2000, 271–272), but one can now see it that it is perfectly compatible with the hermeneutic alternative that I have defended: the point is not that we are just dealing with two different worlds, but that Greek investigative methods had at least some substantial advantages. One may, of course deny this–in line with Heidegger's own practice, it gives a significance to philosophy and its history that few natural scientists would accept—but it is not a relativist view. Similarly consider this, again cited by Lafont:

> The advanced modern science of nature is not a whit more true than the Greek; on the contrary, at most it is more untrue since it is completely caught up in the web of its own methodology, and for all its discoveries, it lets that which is actually the object of these discoveries slip away: namely, the nature of the relation of human beings to it and their place within it.
>
> (Ga 45, 53; cited by Lafont 2000, 272)

The point here is precisely to criticize modern natural science for "letting the object … slip away," and the explanation is exactly what my reading predicts: methodological shortcomings that prevent it from being sufficiently attentive to that object.

The second issue I want to highlight concerns Heidegger's lifelong commitment to what one might call "rampant property pluralism." By this, I mean that he is extremely hostile to programs that reduce certain properties in order to avoid including them in a final ontology. For example, he opposes the standard projectivist stories on which properties such as "toolhood" are reduced to "merely a way of taking" those entities:

> The kind of Being which belongs to these entities is readiness-to-hand. But this characteristic is not to be understood as merely a way of taking them, as if we were talking such "aspects" into the "entities" which we proximally encounter, or as if some world-stuff which is proximally

present-at-hand in itself were "given subjective coloring" in this way ... *Readiness-to-hand is the way in which entities as they are "in themselves" are defined ontologico-categorially.*

(SZ 71, original emphasis)

Similarly, he holds both that individual *Dasein* have causal properties, for example, being a certain mass, and other properties, such as freedom, which must be explained in terms of an entirely different framework (SZ 135, Ga31, 210). While not as important as the preceding issue regarding empiricism and methodology, it is important to bear this in mind when reading Heidegger on the clash between Aristotelian and modern science: there is a sense for him in which both are true, not because of relativism, but because both capture different aspects of the very wide range of properties that entities do, in fact, possess.

Bringing these remarks together, one can now see how Lafont's view actually conflicts with the basic spirit of texts, such as SZ, one of whose central concerns is precisely to highlight and critique cases in which a thinker refuses to revise some initial method, concept, approach, or assumption in the face of the phenomena. For example, Descartes is extensively criticized for imposing a pre-given framework on entities (SZ: 96). Indeed, this is Cartesianism's original sin:

The kind of being which belongs to entities within the world is something which they themselves might have been permitted to present; but Descartes does not let them do so.

(SZ 96)

If Lafont's approach were right, this would not be a criticism but an unavoidable statement of fact.

Conclusion: Relativism as symptom not solution

On the interpretation I have defended, Heidegger was never a relativist. Instead, his position fuses phenomenology, hermeneutics, and Kantianism in a distinctive way, one that gives priority to the process of adjusting our ontology to map the entities and objects we encounter. This process has an inherently circular structure: as we recognize ways in which our understanding is not calibrated to the phenomena, we continually revise that understanding, thus throwing up new "feedback," which in turn forces further revisions (SZ 153). In line with the caveat offered in the introduction, my focus here has been on early Heidegger, but it is very natural to see the shifts in his later work as motivated by a growing fear that the framework of SZ itself failed this test, preventing him from accommodating, from doing justice to phenomena and events from artworks to *physis* itself.

I want to end with one final piece of evidence, and one that might seem to have a particular clarity and directness. As Kusch notes, there is a letter to Löwith in which Heidegger openly identifies himself as a "dogmatic sub-jective relativist" (Kusch 1989, 191). How can my reading handle this?

This is the same document in which Heidegger famously identifies as a "Christian *theologian*" (original emphasis) rather than a philosopher, and it is a complex text. The immediate context is the question of facticity and philosophical method: Heidegger is arguing against a university that he sees as plagued by "fossilized 'intellectualism'" (Heidegger 2007, 101). Philosophically, he defends a position on which "with respect to the things in themselves we are 'absolutely' *objectively rigorous*," but where this rigor arises not from some "fictious non-personality," a view from nowhere, but from an intense personal engagement (Heidegger 2007, 101, original emphasis). We should be wary about trying to reconstruct too precise a view on the basis of an informal letter, but one can see that the contours of Heidegger's position here match those defended above: there is no tension at all between a histor-ical, concrete starting point and an ambition toward accuracy with respect to the things themselves. On the contrary, as I showed above, the two go hand in hand. It is also clear that the relativist label is not one that Heidegger him-self is happy with: he introduces it only on the penultimate page and only to frame the contrast with Löwith, who is identified as an "objective relativist." Heidegger immediately states that the resultant taxonomy is of "no interest to me at all" (Heidegger 2007, 101).

In short, what we see here, exactly as in his published work, is a willingness to recognize some merit in relativism as a crude way of articulating a deeper truth, combined with an insistence on distancing his own philosophy from the term. Ultimately, for Heidegger, relativism is not so much a solution as a symptom, a symptom of the mix of epistemological confusions, tensions, and insights that he takes the tradition to have bequeathed, and which his own hermeneutic model of truth attempts to move beyond.

Abbreviations

References are to the *Gesamtausgabe* edition (Frankfurt: Klostermann, 1975–; abbreviated as Ga), with the exception of SZ, where I use the standard text (Tübingen: Max Niemeyer, 1957). With respect to translations, I have endeavored to stay close to the Macquarrie and Robinson version of SZ on the grounds that it is by far the best known. Where other translations exist, I have typically consulted these but often modified them: the relevant translations are listed below.

SZ *Sein und Zeit* (Tübingen: Niemeyer, 1957); *Being and Time*, translated by J. Macquarrie and E. Robinson (New York: Harper & Row, 1962).

Ga3 *Kant und das Problem der Metaphysik* (1998).

Ga24 *Die Grundprobleme der Phänomenologie* (1997); *Basic Problems of Phenomenology*, trans. A. Hofstadter (Bloomington: Indiana University Press, 1982).

Ga31 *Vom Wesen der menschlichen Freiheit. Einleitung in die Philosophie*
       (1982); *The Essence of Human Freedom*, translated by T. Sadler (London:
       Continuum, 2002).
Ga41 *Die Frage nach dem Ding* (1984).
Ga45 *Grundfragen der Philosophie* (1984).
Ga61 *Phänomenologische Interpretationen zu Aristoteles* (1985).

Notes

1 For a penetrating analysis of the dissertation and its links to the broader debates
   around judgement in the period, see Martin (2006).
2 For the key existing responses to Lafont, see Carman (2002) and Wrathall (2002).
3 The *locus classicus* for the standard reading is Dreyfus (1991); Dreyfus (2005) gives
   a particularly clear exposition of the consequences for language. My own view is
   that Heidegger defends a unique and highly innovative position which is conceptu-
   alist and yet not tacitly propositionalist. To put it another way, his aim is to cash
   conceptuality without appeal to language (for details, see Golob 2014).
4 There are complicated exegetical questions as to what divergence Kant allows for
   non-rational animals or creatures such as angels but for current purposes we can
   simply bracket that and work with a simpler, universalist Kant.
5 This reflects Heidegger's basic understanding of the synthetic a priori as equivalent
   to ontological knowledge. He summarizes the Copernican turn by stating that:

   What Kant wants to say is this: "Not all cognition is ontic and where there is such
   cognition it is made possible only through ontological cognition." (Ga 3, 13)

6 The debate here is closely connected to Tugendhat's influential claim that Heidegger
   robs truth of any normative force (the key text is Tugendhat 1994). Unsurprisingly,
   Lafont is highly sympathetic to such a reading (Lafont 2000, 148). For detailed
   arguments against both Tugendhat and Lafont see Golob (2014, 180–191).
7 I draw here on arguments developed in greater detail in Golob (2014).
8 By extension, Heidegger is positive toward scientists, such as Bohr and Heisenberg,
   who he thinks combine empirical research with this kind of broader methodological and conceptual reflection (Ga 41, 67).

References

Blattner, W. (1999), *Heidegger's Temporal Idealism*, Cambridge: Cambridge University
   Press.
Carman, T. (2002), "Was Heidegger a Linguistic Idealist?" *Inquiry* 45 (2): 205–215.
———— (2003), *Heidegger's Analytic*, Cambridge: Cambridge University Press.
Dilthey, W. (1998), *Introduction to the Human Sciences*, translated by R. Betanzos,
   Princeton : Princeton University Press.
Dreyfus, H. (1991), *Being-in-the-World*, Cambridge, Mass.: MIT Press.
———— (2005), "Overcoming the Myth of the Mental," *Proceedings and Addresses of
   the American Philosophical Association* 79: 47–65.
Foucault, M. (1971), *The Archaeology of Knowledge and the Discourse on Language*,
   translated by A. M. Sheridan Smith. New York: Pantheon Books.

Frede, D. (1993), "The Question of Being: Heidegger's Project," In *The Cambridge Companion to Heidegger*, edited by C. Guignon, Cambridge: Cambridge University Press, 42–70.

Golob, S. (2013), "Heidegger on Kant, Time, and the 'Form' of Intentionality," *British Journal for the History of Philosophy* 21: 345–367.

———(2014), *Heidegger on Concepts, Freedom and Normativity*, Cambridge: Cambridge University Press.

Haugeland, J. (2013), *Dasein Disclosed*, Cambridge Mass.: Harvard University Press.

Heidegger, M. (2007), *Becoming Heidegger: On the Trail of His Early Occasional Writings, 1910–1927*, edited by T. Kisiel and T. Sheehan, Evanston, Ill.: Northwestern University Press.

Husserl, E. (1975), *Logische Untersuchungen. Erster Band: Prolegomena zur reinen Logik*, edited by E. Holenstein, The Hague: Nijhoff.

——— (1997), "Phenomenology and Anthropology," in *Psychological and Transcendental Phenomenology and the Confrontation with Heidegger (1927–1931)*, translated and edited by R. Palmer and T. Sheehan, London: Kluwer, 485–500.

Kisiel, T. (1993), *The Genesis of Being and Time*, Berkeley: University of California Press.

Kusch, M. (1989), *Language as Calculus vs. Language as Universal Medium: A Study in Husserl, Heidegger, and Gadamer*, Dordrecht and London: Kluwer.

Lafont, C. (2000), *Heidegger, Language, and World-Disclosure*, Cambridge: Cambridge University Press.

——— (2002), "Precis of Heidegger, Language and World-Disclosure," *Inquiry* 45 (2): 185–189.

——— (2007), "Heidegger and the Synthetic *A Priori*," In *Transcendental Heidegger*, edited by J. Malpas and S. Crowell, Stanford: Stanford University Press, 105–118.

Martin, W. (2006), *Theories of Judgment: Psychology, Logic, Phenomenology*, Cambridge: Cambridge University Press.

Tugendhat, E. (1994), "Heidegger's Idea of Truth," In *Hermeneutics and Truth*, edited by B. R. Wachterhauser, Evanston, Ill.: Northwestern University Press: 83–97.

Wrathall, M. (2002), "Heidegger, Truth and Reference," *Inquiry* 45 (2): 217–228.

———(2011), *Heidegger and Unconcealment*, Cambridge: Cambridge University Press.

Part IV

Politics

Introduction

Johannes Steizinger

Relativism connects philosophy with politics. Take the current situation. Political motifs are an important issue in philosophical debates about relativism. Relativism is often rejected because of its alleged political consequences. Anti-relativists regard the relativization of beliefs to frameworks as undermining the authority of truth. Allegedly, relativism undermines the distinction between true and false and thereby enables pathological phenomena like "post-truth politics" with its "fake news." Under such a "dictatorship of relativism" (Pope Benedict XVI), there is simply no way of calling out politicians on their blatant lies.

And yet, there is also a relativistic tendency in contemporary political theory. Political struggles by, say, indigenous people or feminists have prompted debates about cultural diversity. The plurality of ways of life and deep disagreements are a permanent challenge to our current polities. In the fierce debate about managing conflicts both within culturally diverse communities and between them, relativism is discussed as a plausible option, albeit mostly critically. Anti-relativists claim that this political context—the contingent connection of relativism with progressive policies such as postcolonialism and feminism—explains the intellectual appeal of relativism in the first place. The critics argue that the adoption of relativism as a philosophical position is motivated more by political correctness than by theoretical insight.

These political aspects of debates on relativism connect the current use of the term with its multi-faceted history, especially in German-speaking philosophy. Issues of *cultural diversity* were already fiercely debated in the second half of the eighteenth century. The discovery of human communities with quite different habits and customs brought out the cultural diversity of humanity. The emphasis on human difference also revealed a downside of the newly established framework of Enlightenment thinking: its universal aspirations suggested ethnocentric treatments of other cultures and justified the atrocities of colonialism. *Vicki A. Spencer* puts Johann Gottfried Herder's (1744–1803) contribution to these eighteenth-century debates in conversation with leading relativists in the social sciences today. She shows that Herder's extensive and concrete studies of different cultures were motivated by an appreciation of cultural diversity. Although Herder emphasized the particularity of human communities in his historical and cross-cultural research, his hermeneutic method is not relativistic. Herder derived "thin" universal principles from his concept of *Humanität* (humanity). Since he used these principles to assess the features of other cultures across the board, he did not adhere to even the weakest forms of relativism discussed in the social sciences today. According to Spencer, the weak relativist has to insist on the non-appraisal of other cultures. Herder was, thus, no relativist, but a "pluralist." Spencer emphasizes that Herder's attempt to balance unity with diversity is the enduring legacy of his methodological and political pluralism (see also Spencer 2012).

The nineteenth century was characterized by deep and rapid changes in all areas of life. Developments such as the rise of the empirical sciences, a new understanding of history, and the progressing secularization changed the intellectual landscape profoundly. As a result, the traditional place and self-understanding of philosophy were challenged. This "identity crisis" of philosophy also fueled the *pejorative use* of the term "relativism." Philosophers such as Wilhelm Windelband or Wilhelm Dilthey presented relativism as a dangerous consequence of the modern spirit. They conceptualized the intellectual changes of modernity as a loss of certainty that threatens the normative foundations of society. Construed as dissolution of fixed values, the rise of empiricism, historicism, and secularization seemed to bring about anarchy and nihilism. It was, of course, the task of philosophy to meet the relativistic challenge of the modern spirit. Hence, the diagnosis of relativism and the philosophical efforts to overcome it were a strategy to re-establish the intellectual authority of philosophy.

As proponents of societal change based on historical insight, Karl Marx (1818–1883) and Friedrich Engels (1820–1895) were, and still are, natural targets of the charge of relativism. *Terrell Carver's* close reading of the drafts of the posthumously compiled *German Ideology* teaches us an important lesson about the politics of relativism. Carver reads the theoretical endeavor of Marx and Engels as an innovative attempt to ground their activist interventions into the politics of the time. (For his reading of Marx as a political activist, see Carver 2018.) He shows that they developed a new concept of knowledge that challenges the philosophical assumptions of both rationalism and empiricism. Marx's and Engels's emphasis on knowledge-making as practical social activity gave rise to a fully socialized and historicized epistemology. This nascent view of knowledge as social practice, practice as politics, and politics as future challenged the philosophical perspective from which the problem of relativism makes sense. According to Carver, relativism only arises from the philosophical prejudice that knowledge must have foundations outside of practical, social activity. From the anti-philosophical outlook of Marx and Engels, relativism is thus no issue at all. In a concrete political context, relativism quickly appears to be nothing but an artificial creation and a whipping boy of philosophers.

The political debates on relativism nevertheless continued and culminated in the early twentieth century. Although the pejorative use of the term "relativism" prevailed, notable exceptions emerged on both ends of the political spectrum: Oswald Spengler (1880–1936) derived his political conservativism from a radical cultural relativism that regarded all kinds of truth as mere expressions of a certain time and place. Benito Mussolini (1883–1945) characterized fascism as a relativistic movement because of its anti-scientism and voluntarism. With his belief in the ultimate authority of pure power, the fascist despised all fixed and stable categories (Mussolini 1921). Hans Kelsen (1881–1973), on the other hand, considered the relativistic denial of absolute truth and values as the prerequisite of liberal democracy. He believed that philosophical absolutism goes hand in hand with political absolutism and is thus the philosophical henchman of despots, dictators, and autocrats.

Yet, not only relativism, but also anti-relativism could motivate quite different political positions. *Johannes Steizinger*'s critical examination of the context of National Socialism (NS) reveals an anti-relativist template that was used by Nazi philosophers, academic philosophers, and Nazi critics alike. Here, the charge of relativism was an important polemical motif to discredit the philosophical and/or political enemy. This polemic use rested on a specific understanding of relativism: Relativism was considered as a fundamental problem that has to be overcome. It was thus depicted as a vague threat that endangers not only philosophy proper, but society and life in general. Moreover, it was always the same strands of nineteenth-century philosophy that were found guilty of having caused this problem and the subsequent "crisis" of the modern spirit. Steizinger presents the politics of relativism in the context of NS as typical case of the pejorative use of the term. The analysis of the actual debate on relativism during NS also shows its enduring connection with the issue of cultural diversity. The relativistic tendency of NS arose from a radical critique of universal concepts of humanity. Nazi philosophers insisted on the particular identity of *Völker* (people) and regarded universal claims as purely ideological mechanisms. They considered the ethnic community as only source of normative authority. Yet, they did not advocate tolerance of other ways of life or neutrality when being confronted with different worldviews. Nazi ideology was a *racist* anthropology. Nazi philosophers believed in an objective hierarchy of races and attempted to justify their ranking. The conviction that there is a *Herrenrasse* (master race) and that its superiority can be demonstrated is the non-relativistic core of Nazi ideology. Although Nazi philosophers presented their view as overcoming the opposition between absolutism and relativism, they could never resolve the tensions between its relativistic tendency and its anti-relativistic assumption.

References

Carver, T. (2017), *Marx*, Cambridge: Polity Press.

Mussolini, B. (1921), "Nel solco delle grandi filosofie: relativismo e fascism," *Il Popolo d'Italia*, November 22.

Spencer, V. A. (2012), *Herder's Political Thought: A Study of Language, Culture and Community*, Toronto: University of Toronto Press.

12 Unity and diversity

Herder, relativism, and pluralism

Vicki A. Spencer

Introduction

It is now well established that Johann Gottfried Herder (1744–1803) never entirely abandoned the Enlightenment with its commitment to universal values. Nonetheless, it remains the practice of some Herder scholars to employ the term *relativism*—a philosophical position that rejects universal normative claims—to describe his thought. Sonia Sikka (2011), for example, has recently maintained that Herder is a relativist in relation to virtue and happiness. In contradistinction, I have argued elsewhere (Spencer 2012, 1998) that the most coherent position in modern philosophical terms to attribute to Herder's moral theory is a pluralist one that is characterized by a non-dualistic commitment to universal values and the particular.[1]

It is quite possible, nonetheless, for Herder to have been a moral pluralist while being a relativist in methodological terms. It is important to acknowledge that it is entirely plausible for a thinker to be a relativist in one domain and not others. One might think that artistic taste, for instance, is largely determined by, and therefore relative to, our cultural influences while insisting upon the existence of universal human rights in the moral domain. Or one might consider science produces objective and universal truths, while still being a moral relativist.

In this chapter, I focus on the extent to which Herder might reasonably be categorized as a methodological relativist. Here, his historicism takes center-stage, and I show considerable textual evidence exists in his work to inspire a relativist methodology for historical and cross-cultural research. Yet, despite a good deal of common ground existing between Herder's work and leading relativist thinkers in the social sciences today, I argue that Herder's work ultimately fails to adhere to modern standards for a relativist methodology, even in its weakest form. While the weak relativist passes no judgment in historical or cross-cultural studies, Herder's historical and aesthetic works are imbued with moral evaluations based on his commitment to a set of general, albeit "thin," universal principles.[2] Thus, to extract a relativist methodology from Herder's work is to concentrate one-sidedly on certain parts of his work and results in a distortion of the approach he employs in his own cross-cultural and historical studies. In accord with contemporary social and political theory, that approach is best captured, in my view, by the term "pluralism."

Cultural and methodological relativism

David Barnes and Barry Bloor (1982) indicate the simple starting point of relativism is the recognition that beliefs on a certain topic vary. But this empirical fact is hardly denied by the universalist either. Immanuel Kant (1991), for instance, was just as aware of the diversity of beliefs that human beings held in the eighteenth century as Herder was. He thought we ought to adhere to the universal moral law, but he was not so naïve to believe that we do so. Recognition of diversity is thus a necessary, but by no means sufficient condition, for a relativist position.

The kind of diversity in question is pertinent, too. As Hannah Arendt (1958) indicates, pluralism in a descriptive sense is a fundamental hallmark of the human condition at the level of the individual. But relativism is most commonly employed in relation to diversity between collectivities. Unlike subjectivism, in which values are considered a matter of personal and individual taste and opinion, relativism possesses an objective component as values are typically considered context-bound to one's culture and/or one's schematic framework. Due to his sustained interest in cultural diversity, it is cultural relativism that is typically attributed to Herder's thought.

I take two key commitments as central to the cultural relativist position. First, that the truth or falsity of a belief or the validity of a practice is relative to the cultural context in which that belief is held or the practice is performed. For the cultural relativist, there is no sense in which some beliefs are really rational, true, or good, because there are no context-free, transcultural, or transhistorical norms of rationality, rightness, or goodness as distinct from those that are locally accepted within the context of a particular culture.

Here, an important distinction is often made. Weaker versions of the theory do not deny that some of our convictions are shared by other cultures, even if they reject universal normative claims; otherwise, it would be impossible for us to communicate with their members or to understand them. Although strong relativism treats cultures as if they are virtually self-contained systems, weaker versions acknowledge certain factual universals due to our common human existence, such as the reality that human beings are linguistic creatures, and in common they require shelter, food and the fulfillment of other basic needs. In other words, weaker relativists cannot be accused of treating different cultures as if their members belong to an entirely different and alien species.[3]

The second central component of any relativist position is a symmetry or equivalence postulate. Again, there are stronger and weaker versions. The strong version is evident in the common relativist view that the beliefs and practices of all cultures are equally valid. This formulation is, however, open to the immediate objection that if no transcultural and transhistorical values exist, we are unable to make such a comparative judgment. We cannot possibly know whether something is better or worse or equal in value. Charles Taylor (1992, 68–73) makes the additional point that to make such an *a priori* judgment before engaging in any empirical investigation of the culture in question amounts to a patronizing, albeit well-intentioned, ethnocentricity based on a desire to be tolerant. Yet the universal demand for toleration toward difference that often follows from this strong equivalence postulate leads to an additional contradiction for a theory that denies the existence of universal values. Relativism, in this strong form, is thus routinely dismissed as incoherent (Williams 1985, 1972; Rachels 2007).

Not all relativists, however, make this strong claim. Barry Barnes and David Bloor (1982) instead maintain that "all beliefs are on a par with one another with respect to the cause of their credibility" (23). This postulate is the basis for their methodological relativism whereby they advocate "that the incidence of all beliefs without exception calls for empirical investigation and must be accounted for by finding the specific, local causes for their credibility" (Barnes and Bloor 1982, 23). The role of the social scientist is to explain the basis for certain beliefs and the form of reasoning that exists to justify them without any regard to whether they are true or good. The investigator may have her moral perspective on those beliefs, but they are inappropriate in social science research. Criticism of others' beliefs by the social science researcher is ruled out because the investigator believes her preferences and evaluations are just as context-bound and relative to her culture as those under investigation (Barnes and Bloor 1982, 27). There might be certain universal facts about human beings and their basic needs, but there are no universal values. Thus the belief systems or individual practices of different cultures cannot be ranked. They are neither equally valid nor false; non-appraisal, rather than equality, is all that is required in this weaker variant of the equivalence postulate (Kusch 2016).

Herder and relativism

There are a number of significant overlaps between the weaker relativist position proposed by Barnes and Bloor for social science research and the historicism Herder developed. Herder most famously elaborated the basis of his historicist approach in his early essay *Auch eine Philosophie der Geschichte* (*This Too a Philosophy of History*, 1774). It is particularly well-known for his spirited critique of Enlightenment philosophers including Hume, Voltaire, Robertson, and Iselin for their employment of contemporary European standards to judge the relative civilization or backwardness of other cultures. Herder (2002b, 307–308, 1994, 51) is highly disparaging of those philosophers who portray historical progress as a steady, linear progression from superstition and ignorance toward a superior, enlightened life of the present. In the aesthetic domain, he is equally critical of Johann Winkelmann for having elevated Greek artistic standards to a universal status that he then employed to analyze the merits and demerits of the artistic achievements of other cultures (Herder 2002b, 283, 1994, 23).

Herder insists that the ranking of cultures is invalid. It is pointless to criticize a people for being unlike another that the researcher might prefer when the historical and environmental conditions for each differ so vastly. It is simply not possible, for example, for a community of shepherds to act in the same way as the Romans who "were able to be as no other nation, to do what *no one does in imitation*—they were *Romans*" (Herder 2002b, 295, 1994, 37). In a community of shepherds, one finds virtues that the Romans or Spartans did not possess and vices among the Romans and Spartans that are non-existent in farming and shepherding communities. All cultures and eras, he asserts, embody negative and positive features (Herder 2002b, 294–295, 1994, 36–37).[4]

Unbalanced praise that fails to pay equal attention to the shortcomings of any given society produces an inaccurate depiction of the society under a researcher's investigation. Herder (2002b, 276–277, 1994, 15–16) acknowledges the important role our prejudices play in situating us, and he does not naively believe that we can dispense with them entirely.[5] Indeed, he is certainly incapable of hiding his enthusiasm for the flourishing of the arts and sciences in ancient Athens. However, it is possible for us to develop a reflexive awareness of our prejudices and by consciously directing our attention also toward a society's shortcomings, we can arrive at the more balanced perspective that Herder considers necessary for historical understanding. For it is only "[w]hen the exaggerated reverence will have been blunted, the factionalism with which each person *cuddles* his people as a Pandora sufficiently brought into balance—you *Greeks* and *Romans*, then we will know you and classify you!" (Herder 2002b, 341, 1994, 89). Thus, he also points out their flaws.

Central to the task of the historian is to explain the causes and rationale for historical events and the practices of different societies, not to engage in exaggerated praise or criticism. In *Auch eine Philosophie*, Herder applies the hermeneutic reading of texts he developed in his earlier work to the study of history.[6] In his 1768 commentary *Über Thomas Abbts Schriften* (*On Thomas Abbt's Writings*), he describes, "the man I want is the *explainer* who defines the borders of an author's past world, own time, and the world of posterity— what the first supplied to him, how the second helped or harmed him, how the third developed his work" (2002b, 173, 1993b, 580). Reading is both an imaginative and interpretative process, but to understand a text, we need to engage with it dialogically and attempt to understand the author's way of thinking with all its uniqueness; we need to try to see it from the author's perspective. The achievement of this objective necessitates that we learn about the historical and cultural context in which the author wrote, for Herder considers the best commentator on an author is one "who does not *modify* him to accord with his own century, but explains him in all the nuances of his time" (2002b, 172, 1993b, 579).

The historical method he develops to assist in this process is based on an empathetic identification with one's subject: "one would have first to *sympathize* with the nation, in order to feel a *single one* of its *inclinations* or *actions all together*" (Herder 2002b, 292, 1994, 33). Data collection on a society's constitutional rules and history is vital. But Herder identifies our imaginative and sensuous powers (2002b, 292, 1994, 33–34) as the keys to opening a pathway toward greater understanding when he recommends that it is necessary to "go into the age, into the clime, the whole history, feel yourself into everything." It is only then we can come to understand our subjects in their own terms and dispense with the habit of comparing everything to our own likes and dislikes. In his commentary on one-sided denunciations of the Middle Ages as barbaric, irrational, and superstitious, he states, "I would rather do anything than defend the eternal peoples' migrations and devastations, vassal-wars and fights, monks' armies, pilgrimages, crusades—I would only like to explain them" (2002b, 309, 1994, 53). With the use of his empathetic approach, he discovers that people during the Middle Ages constantly struggled against their shortcomings and strove for improvement. The era cannot therefore legitimately be characterized as entirely barbaric (Herder 2002b, 309, 1994, 53).

In his most important mature historical work, *Ideen zur Philosophie der Geschichte der Menschheit* (*Ideas for a Philosophy of History of Humanity,* 1784–1791), Herder repeats many of the same ideas. In a passage in *Auch eine Philosophie*, which is often taken as strong evidence of relativism in Herder's thought, he states, "Each nation has its *center* of happiness *in itself*, like every sphere has its center of gravity!" (Herder 2002a, 297, 1994, 39). Similarly, in the *Ideen*, he insists, "It would be the most stupid vanity to imagine that all the inhabitants of the world must be Europeans to live happily" (2002a, 298, 1800, 219). The time and place into which one is born are arbitrary. So, if a European happened to be born into a non-European community, he or she would develop non-European standards. Differences in these standards are no concrete grounds to assume, at the outset, the inferiority of non-European values and practices that are developed in very different circumstances. "He who placed them here and us there," he writes, "undoubtedly gave them the equal right to the enjoyment of life." Happiness is an "internal disposition" that is influenced by the culture we are immersed in from birth (Herder 2002a, 298, 1800, 218–219).

In the preface to the *Ideen*, Herder develops a neutral definition of culture that he distinguishes from the notion of "a cultivated people" he emphatically rejects. In a clear refutation of Kant's distinction between a raw and cultivated state, Herder sees no justification in granting preeminence to the "cultivated" few over the majority (Kant 1991, 48–49).[7] He instead employs the term *Cultur* in the modern anthropological sense to indicate a particular way of life of a period, people, or group when he asks, "Which people on earth is there that does not have a common culture [*Cultur*]?" (Herder 2002a, 9–10). He thus departs from both the use of the term in England, France, and Germany in the eighteenth and nineteenth centuries to mean civilization and its common usage in more recent times to refer to intellectual and artistic activities.[8] For Herder, political institutions are as much a part of a people's culture as are its artistic pursuits.

While retaining his empathetic and hermeneutic approach that distinguishes his research methodology from positivism, in the *Ideen* he emphasizes the need for historians to adopt the kind of neutrality characteristic of the natural sciences. "In the narration of history," Herder (2002a, 522, 1800, 392) writes, "one will therefore seek the strictest truth and the most complete connection in its version and assessments, and never strive to explain a thing, which is or happens, by a thing that it is not." No scientist would, for example, attempt to judge a sloth for its failure to perform the actions of an elephant. Equivalent comparisons are therefore equally out of place in historical research (Herder 2002a, 466–467). Even when evident advances exist in certain fields such as the sciences—and Herder (2002a, 604–606) admits that the progress in geology and trade in the eighteenth century, compared to ancient times, is undeniable—"it does no justice, to no people on earth, to judge them by a foreign standard of science" (Herder 2002a, 502, 1800, 377). Not only has contemporary science benefitted from, and built on, earlier discoveries, but there is no historical gain in such judgments as they merely prevent us from coming to understand and appreciate the scientific insights the ancients did have.

To summarize, so far, considerable evidence can be gathered to justify the attribution of a relativist methodology to Herder's work if not in a strong form, at least, in a weak version. First, like Barnes and Bloor, Herder insists on the necessity for the researcher to conduct detailed empirical investigations that seek to explain the local basis of credibility for the particular beliefs and practices held by a community. Second, the ranking of different cultures according to some scale of civilization or rationality along with the judging of one culture by the particular standards of another are deemed invalid. Third, Herder considers that each culture has its own unique conception of happiness and that standards of virtue differ between cultures. Fourth, Herder makes no claim that all cultures are equally valid. In the sciences, he considers that progress evidently occurs, but such judgments are nonetheless inappropriate in historical terms.[9] Thus, Herder's historicism would appear to conform to a doctrine of non-appraisal rather than equality in accord with the weak relativist methodology Barnes and Bloor propose. In the remainder of this chapter, I will nonetheless show that such a conclusion is premature and ignores the way Herder employs his methodology in practice.

Beyond relativism

Particularly pertinent to note in his early *Auch eine Philosophie der Geschichte* is Herder's desire for the historian to present as balanced an appraisal as possible of the negative and positive features in any given society. Yet to set the historian the task of pointing out the virtues and shortcomings that Herder believes exist in all complex human societies means the historian needs to engage in evaluations and judgments. It would be possible for the historian merely to explain what either the Romans or Spartans considered were virtuous forms of behavior or vices within the context of their times and the reasons for their beliefs as Barnes and Bloor recommend. Herder, however, goes considerably beyond presenting the perspective of the ancients when he writes, "About noble Spartans there dwell inhumanly treated *Helots*. The Roman *victor* dyed with *red dye of the gods* is invisibly also *daubed* with *blood; plunder, wickedness*, and *lusts* surround this chariot; before him goes *oppression*, in his train follows *misery* and *poverty*" (Herder 2002b, 295, 1994, 37). Of course, it is possible that some Spartans did think their treatment of helots was inhumane. Herder acknowledges the great diversity of views that exist within any given society and the difficulty of generalizations when it comes to entire cultures, but any such criticism that might have existed among the Spartans was clearly not the dominant perspective. Similarly, the Romans typically did not consider their victories in their expansion of their empire to be wicked or oppressive in the negative sense Herder employs these terms in the above quotation.

The same method of seeking to present a balanced appraisal of the positives and negatives in any given society is central to his analysis of the numerous societies he studies in the *Ideen*. To take just one example, Herder dutifully presents the positive features of Indian society that includes, to his mind, Hindu religious tolerance, something the Hindus also consider a virtue along with their caste system. Similarly, at least for the majority of Hindus, their treatment of "Untouchables" was entirely acceptable within their own schematic framework. Yet Herder, by contrast, is highly critical of the contempt with which "Untouchables" were treated and of the Hindu caste system, in general, which he considers detrimental to the advancement of the arts (Herder 2002a, 415–416). Thus, while his historical studies show what societies consider to be virtuous differs, something is not virtuous from Herder's perspective simply because a society thinks it is. He also finds the Hindu practice of burning wives on the funeral pyres of their dead husbands lacks any legitimate justification:

> The burning of wives on the funeral pyres of their husband may be reckoned among the barbarous consequences of this doctrine: for to whatever cause it owes its first introduction, whether it entered the course of custom either as an emulation of some great minds or as punishment: the doctrine of the Brahmins of a future state has unquestionably ennobled the unnatural practice and inspired the poor victim to encounter death. No doubt this cruel practice renders the life of the husband dearer to the wife, as she thus becomes inseparable from him even in death, and cannot remain behind him without disgrace; but is this gain worth the sacrifice, when only tacit custom gives it the force of law?
>
> (Herder 2002a, 416–417, 1800, 309)

Herder clearly attempts to provide a rationale for this practice in terms of the loyalty it engenders in a wife for her husband, but as a researcher he does not confine his analysis to mere explanation.

Herder's examination in the *Ideen* of the practice of cannibalism among some New Zealand Māori, and "the Eskimo"[10] practice of leaving elderly parents in the snow to die are further instructive. He considers that all human beings possess *Humanität, Vernunft und Sprache* (humanity, reason, and language), but one's *Humanität* can be overridden by other considerations such as greed, the desire for power, or the necessity for sheer survival, or it can simply be undeveloped just as one's capacity for rational critical thinking can be (Herder 2002a, 345). As part of the human species, New Zealand Māori and "Eskimos" possess *Humanität*, reason, and language, and it is due to these universals that it is possible for us to understand the reasons for their actions. Herder (2002a, 345, 1800, 255) explains: "No cannibal devours his brothers and children; their inhumane practice is a savage right of war, to nourish their valor, and terrify their enemies. It is no more or less than a gross political rationale." He attributes "the Eskimo practice" to "lamentable necessity" for the sake of the group's survival, not due to a lack either of love for their parents or morality (Herder 2002a, 345, 1800, 255).

Herder thus shows in accord with Barnes and Bloor's weak relativist methodology that these cultural practices, which are alien to our own experience, possess an intelligible rationale within their own context. Māori did not choose to eat their loved ones, nor did they apply the practice arbitrarily. It was a rule-governed practice of war that they performed on their captured enemies, and they expected to be subjected to the same treatment if they were captured. Moreover, despite its foreign nature, we can relate to and understand it because our own society is replete with examples whereby our own *Humanität* is overcome in the interests of power politics. To err in these ways is entirely human. The only distinction from Herder's perspective is that Europeans "overpower their *Humanität* in some other respects" (Herder 2002a, 345, 1800, 255).

However, despite cannibalism being intelligible to us due to its common traits with our own practices, Herder considers it an "inhumane" practice. He thus goes beyond the relativist methodology recommended by Barnes and Bloor by evaluating others' cultural practices according to his own criteria. One could respond that there is an exceptionally weak equivalence postulate evident in Herder's historical and cross-cultural methodology as he considers all cultural practices to be intelligible by possessing a certain rationale in terms of instrumental reason, but that is a necessary premise to conduct any research in the humanities and social sciences. In terms of the validity of attributing a weak relativist methodology to Herder's work, the relevant point is that he does not adhere to a non-appraisal equivalence postulate.

In the aesthetic domain, Herder also goes beyond merely explaining the rationale of a stylistic choice by an author in a particular culture as he recommends in the above quotations from his *Über Thomas Abbts Schriften*. The first crucial step is to explain the reasons for an author's choices. It also remains necessary not to expect authors to adopt the cultural standards of a different place and time or to criticize them for not doing so. As he maintains in his essay *Shakespeare* (1773):

In Greece the drama developed in a way that it could not in the north. In Greece it was what it can never be in the north. In the north it is not and cannot be what it was in Greece. Thus Sophocles' drama and Shakespeare's drama are two things that in a certain respect have scarcely their name in common.

(Herder 2006, 292, 1993b, 499–500)

Yet Herder draws from this difference that it is necessary for artists not to engage in an attempt to imitate the artistic norms of a different place and era, something he considers has universal relevance. This principle then forms the basis for his criticism of French classicism for its adoption of Greek forms. He acknowledges its positive elements by noting its "beautiful *scenes, dialogues, verses,* and *rhymes* with their *measure, decorum,* and *brilliance*" (Herder 2006, 501, 1993b, 503), but he adds, "for all that, there is still the oppressive, inescapable feeling that 'this is no Greek Tragedy! This is no Greek drama in its purpose, effect, kind, and nature!' ... even if we admit they do keep to these rules, French drama is not the same as Greek drama" (Herder 2006, 295, 1993b, 504).

As Rachel Zuckert (2015) argues, Herder evaluates whether or not artistic works contribute to human flourishing. Key to that flourishing is authenticity to one's own culture, time, and place. Shakespeare did not have a Greek chorus available to him as Sophocles did. But, by drawing on the puppet plays and historical dramas that existed in Elizabethan England, he was able to create a new dramatic form as unique and vibrant as Sophocles' plays. Despite all their differences, Sophocles and Shakespeare are "inwardly" alike as they both looked to their own culture for inspiration and resources. Underlying the great diversity within human history, we discover a unity: "we see that the whole world is merely the body belonging to this great spirit: all the scenes of Nature are the limbs of this body, just as every character and way of thinking is a feature of this spirit—and we might call the whole by the name of Spinoza's vast God: 'Pan! Universum!'" (Herder 2006, 303, 1993b, 515). In the mythologies of different peoples, too, Herder maintains that "uniformity amidst diversity" is evident (Herder 1993a, 80). Thus, despite Herder's commitment to diversity and the particular, he finds it possible to pinpoint certain principles and goods that assist the flourishing of human beings in whatever era or place they reside.

Toward pluralism: Unity-in-diversity

Herder's sustained attention to particularity means his thought needs to be distinguished from the universalism of the majority of his contemporaries. He did not, for example, think there is simply one way to live the good life, and he rejected the idea of the existence of one political constitution suitable for all times and places. There are many valuable and worthwhile forms of life. But, as we have seen, his historical and cross-cultural research simultaneously showed him there are certain general principles that are far more conducive to human self-realization and fulfillment than others. The purpose of Herder's historical work, moreover, is precisely to discover both the manifold diversity that characterizes humanity and the common lessons we can draw from it. Like Rousseau, he did not, for example, believe that the emerging states in the late eighteenth century could adopt the participatory democracy of ancient Athens. Nor did he want to adopt the negative features of Athenian democracy with its exclusive citizenry and institution of slavery (Herder 2002a, 499). But the greater freedoms that Athenian male citizens enjoyed served as the inspiration for Herder's republicanism.[11] We should not attempt to imitate blindly the practices in other cultures; but he emphasized that we can, and should, learn from them by adapting and reinterpreting the spirit of those practices that assist human flourishing to suit our own time and place (Herder 1991, 226, 699–700).

The principles that form the basis for Herder's criticisms in his early work become far more explicit in his mature work with the development of his concept of *Humanität* in the *Ideen*, which he further refines in his *Briefe zu Beförderung der Humanität* (*Letters for the Advancement of Humanity*, 1793–1797). He writes in the *Ideen*,

> I wish I could grasp everything in the word *Humanität*, that I have so far said about the noble formation of the person to reason and freedom, to finer senses and drives, to the gentlest yet strongest constitution, to the discharge and rule of the earth: because there is no nobler words for the person's destination than that which expresses himself, in whom the image of the creator of our earth lives imprinted as visibly as it can be here.
>
> (Herder 2002a, 142–143)

Herder (1991, 148) later clarifies in the *Briefe* that his concept should not be reduced to either "humaneness" or an "unqualified love for humankind." Nor is it another word for humankind or humanitarianism, although humaneness is an important element in its attainment (Frazer 2010, 162; Adler 2009, 1994, 63–65). He refers to human love, dignity, justice, sympathy, compassion, reason, equity, goodness, and even beauty, at times, to describe *Humanität*, but he is aware that none of these terms fully captures its meaning (Herder 2002a, 144–146, 600 f., 1991, 147–148). *Humanität* as a normative concept in Herder's thought is best understood as human nature *par excellence*.

The factual universals Herder considers characterize the human species include our erect stature that gives us greater freedom over other species and our capacity for language. He also believes that people are "designed to be a mild peaceful creature" (Herder 2002a, 127, 1800, 86); it is only oppression that distorts this nature. Yet it does not follow from the fact that people are peaceful that they ought to be (Crowder 2007). It is for this reason that an acknowledgment of the existence of factual universals is consistent with a weak relativist position. However, unlike the weak relativist, Herder goes beyond the acceptance of factual universals among human beings with his normative commitment to the following ten principles he considers essential for human beings to live a good life:

1. *Humanität* is achieved through the authentic self-realization of each individual.
2. This self-realization requires recognition that we are situated selves within a particular cultural and historical context.
3. Harming and oppressing others violates this ultimate end.
4. Along with the avoidance of radical harm, the realization of our *Humanität* requires us to live in peace (Herder 2002a, 143).
5. Our destructive powers need to be controlled by reason in relation to equity and goodness (Herder 1991, 128).
6. The good life requires adherence to the principle of fairness: "*Do not onto others what you would not wish them to do unto you; what you expect others to do to you, do unto them too*" (Herder 2002a, 143).[12]
7. In addition to tolerance, we need to develop a transnational empathy (Herder 1991, 723–724).
8. We have a duty to help others and, in particular, the weak and oppressed (Herder 1991, 130–131).
9. Human beings not only require the freedom to choose and therefore err so they can learn—they should choose (Herder 2002a, 135).
10. Every individual and community needs to interpret these general normative principles for themselves in relation to the social realities that confront them.[13]

Adherence to merely one of these universally applicable principles might not result in the realization of our *Humanität*. In the case of cannibalism, Herder shows that the Māori who performed this practice adhered to his principle of fairness, and yet he judges it an inhumane practice because of the radical harm it inflicted on those captured in war. Since every era and culture possesses negative along with positive features, Herder never depicts an entire era or culture as the embodiment of *Humanität*. Nonetheless, the realization of the *Humanität* we all possess is evident in particular practices at different points throughout history and in the actions of individuals. He highlights the Quakers' opposition to slavery along with the work of Las Casas, Fénelon, and the Abbé St-Pierre in promoting justice and peace as manifestations of *Humanität* (Herder 1991, 148). His concept thus operates as a *telos* toward which individuals ought to strive, but its realization does not mark the endpoint of history or some future utopia. Due to Herder's grasp of human fallibility, he recognizes that our *Humanität* can be realized at one moment in history, only to be lost in the next; the wise and good, he indicates, are often overcome by the foolish and wicked (2002a, 168).

Just as Shakespeare's and Sophocles' dramas cannot be ranked because they each represent moments of authentic genius, when an activity, virtue, or individual action exemplifies *Humanität* they are incommensurable. Thus, Herder (2002a, 538, 1800, 452) maintains that the "culture of the beauty of ancient

Greece, particularly in Athens" with its art and politics; the "virtue" of the Spartans and Romans in their dedication to their republics; the "refined purity and quiet labor and endurance of the Hindus"; and "the spirit of navigation and commercial diligence of the Phoenicians" are "almost non-comparable." Yet the non-appraisal equivalence postulate characteristic of weak relativism does not capture Herder's methodology. He has evidently evaluated these practices as praiseworthy, compared to the "inhumane" practice of cannibalism, or the "barbaric" treatment of helots and colonies by the Athenians, or hereditary government that he dismisses in the *Ideen* as one "of the darkest formulations of the human language" (Herder 2002a, 332). Communal traditions, he further writes, lose their validity when they hinder "the thinking faculty in politics and education, and prevent all progress of human reason and improvement according to new circumstances and times" (Herder 2002a, 470, 1800, 352).

Herder's application of these universal values in his historical and cross-cultural studies, combined with his attention to diversity, has often led to the conclusion that his work is replete with contradiction.[14] However, this conclusion was largely the consequence of a philosophical framework that insisted either there is some fixed foundation we can appeal to in order to ascertain the truth in accord with a strong universalism, or we find ourselves in the midst of an intellectual and moral chaos that was attributed to relativism (Bernstein 1983). Over the past thirty years, with the pluralist turn in political philosophy, this either/or has largely been discredited. Yet this dichotomy was always inappropriately applied to a holistic thinker who rejects mutually exclusive categories in the interests of capturing the complexity of human existence.

That he sees happiness as an "internal disposition" influenced by one's cultural framework is also no more evidence that he is a relativist than Kant's view that happiness is subjective can be used to argue that Kant is a subjectivist.[15] Herder's work is characterized by a sustained attempt to capture the universal *and* the particular. His universalism is not a strong or "thick" version, so he does not consider that there is only one definitive way to live the good life. However, to employ Michael Walzer's terminology, he adheres to a "thin" version (Walzer 1994) that acts as a *telos* toward which we ought to strive and that assumes multiple concrete forms throughout history. Herder then employs this thin universalism to analyze the extent to which individual human practices and actions in history contribute to or hinder human welfare. It nonetheless remains invalid for the social scientist to rank entire cultures or eras, since such a practice tends to produce unbalanced generalizations that hinder our ability to understand others and discover the manifold ways in which *Humanität* can be manifest.

Conclusion

In modern philosophical terms, Herder's historical and cross-cultural work is most appropriately seen to exemplify a pluralist methodology. He recommends the kind of neutrality found in the natural sciences for the collection of data from different cultures, combined with the use of empathy to assist us to envision our subjects' perspectives and the rationale for their behavior. However, unlike Barnes and Bloor, he does not confine the researcher to mere explanation. The task of the historian is to present the virtues and shortcomings of different societies in a balanced way. Herder thus urges us to approach justly the study of those living in very different conditions by considering their historical context and seeking to understand the great diversity characterizing humankind in its own terms. But to determine the value as well as the shortcomings each society possesses, Herder then evaluates their individual practices and institutions according to the thin normative commitments encapsulated in his later works in his concept of *Humanität*.

Notes

1 For a discussion of the pluralist position in modern philosophy see Rachels 2007, ch. 3.
2 For the distinction between a "thick" and "thin" universal morality, see Walzer 1994.
3 Previously, I did not make this distinction, as it was usually a strong version of relativism that was attributed to Herder's thought (see Spencer 1998, 2012).
4 For a fuller discussion of Herder's recognition of diversity *within* cultures and his rejection of Leibniz's theory of monads as self-contained entities, see Spencer 2012, 70–75, 88–89, 2007.
5 For a discussion of Herder's approach to prejudice, see Menze 2002, Irmcher 1977, 532.
6 For a recent discussion of Herder's hermeneutics, see Gjesdal 2017.
7 On this distinction between Herder and Kant, see Heinz 1996, 142–146.
8 The only full translation of Herder's *Ideen* is T. Churchill's *Outlines of a Philosophy of History of Man,* published in 1800, in which he mistakenly translates Herder's use of the term *Cultur* as "civilization." Thus, I have adopted the practice of citing Churchill's translation to assist English readers, but I have liberally adapted his translations in light of the original German. When I have not done so, the translation is my own. See Raymond Williams (1983, 87–90) on Herder taking a new direction with his usage of the term *Cultur*.
9 Due to the influence of Thomas Kuhn on the work of Barnes and Bloor, they also apply their relativism to science (Barnes 2011).
10 I acknowledge that in Canada, the term *Eskimo* is considered an insult, and the term *Inuit* is preferred. However, in Alaska the term also refers to the Yup'ik, who do not consider themselves Inuit. Since it is unclear whether Herder was referring to the Inuit or both peoples, I employ Herder's term in inverted commas.
11 On Herder's republicanism, see Spencer 2012, ch. 6 and Barnard 1965.
12 For a detailed discussion of this aspect of his concept see Adler 2009, 108–111.
13 This list is a modification of my previous one, as it only includes Herder's normative commitments that distinguish his position from a weak relativism. For a more extensive discussion of his concept of *Humanität,* see Spencer 2012, 112–118.
14 For the classic statement, see Meinecke 1972, 354, 366–368.
15 For an argument that Herder is a strong relativist when it comes to measuring happiness and, indeed, individual virtue, see Sikka 2011, 33–39.

References

Adler, H. (1994), "Johann Gottfried Herder's Concept of Humanity," *Studies in Eighteenth-Century Culture* 23: 55–74.

—— (2009), "Herder's Concept of Humanität," in *A Companion to the Works of Johann Gottfried Herder*, edited by H. Adler and W. Koepke, Rochester, N.Y.: Camden House, 331–350.

Arendt, H. (1958), *The Human Condition*, Chicago: University of Chicago Press.

Barnard, F. M. (1965), *Herder's Social and Political Thought: From Enlightenment to Nationalism*, Oxford: Clarendon Press.

Barnes, B. (2011), "Relativism as a Completion of the Scientific Project," in *The Problem of Relativism in the Sociology of (Scientific) Knowledge*, edited by R. Schantz and M. Seidel, Frankfurt: Ontos Verlag, 23–39.

Barnes, B. and D. Bloor (1982), "Relativism, Rationalism, and the Sociology of Knowledge," in *Rationality and Relativism*, edited by M. Hollis and S. Lukes, Oxford: Clarendon Press, 21–47.

Bernstein, R. J. (1983), *Beyond Objectivism and Relativism: Science, Hermeneutics, and Praxis*, Philadelphia: University of Pennsylvania Press.

Crowder, G. (2007), "Value Pluralism and Liberalism: Berlin and Beyond," in *The One and the Many: Reading Isaiah Berlin*, edited by G. Crowder and H. Henry, New York: Prometheus Books, 207–230.

Frazer, M. L. (2010), *The Enlightenment of Sympathy: Justice and the Moral Sentiments in the Eighteenth Century and Today*, Oxford: Oxford University Press.

Gjesdal, K. (2017), "Human Nature and Human Science: Herder's Hermeneutics," in *Herder: Philosophy and Anthropology*, edited by A. Waldow and N. DeSouza, Oxford: Oxford University Press, 166–184.

Heinz, M. (1996), "Kulturtheorie der Aufklärung: Herder und Kant," in *Nationen und Kulturen: Zum 250. Geburtstag Johann Gottfried Herders*, edited by R. Otto, Würzburg: Königshausen and Neumann, 139–152.

Herder, J. G. (1800), *Outlines of a Philosophy of History of Man*, translated by T. Churchill, London: Printed for J. Johnson by Luke Hansard.

—— (1991), *Briefe zu Beförderung der Humanität*, edited by H. D. Irmscher, vol. 7 of Herder, *Werke in zehn Bänden*, ed. G. Arnold et al., Frankfurt: Deutscher Klassiker Verlag.

—— (1993a), "Fragment of an Essay on Mythology," in Herder, *Against Pure Reason: Writings on Religion, Language, and History*, translated and edited by M. Bunge, Eugene: Wipf and Stock, 80–83.

—— (1993b), *Schriften zur Ästhetik und Literatur, 1767–1781*, edited by G. E. Grimm, Frankfurt: Deutscher Klassiker Verlag.

—— (1994), *Schriften zu Philosophie, Literatur, Kunst und Altertum, 1774–1787*, edited by J. Brummack and M. Bollacher, Frankfurt: Deutscher Klassiker Verlag.

—— (2002a), *Ideen zur Philosophie der Geschichte der Menschheit*, in Herder, *Werke*, vol. 3, bk. 1, edited by W. Proß, Munich and Vienna: Carl Hanser Verlag.

—— (2002b), *Philosophical Writings*, translated and edited by M. Forster, Cambridge: Cambridge University Press.

—— (2006), *Selected Writings on Aesthetics*, translated and edited by G. Moore, Princeton: Princeton University Press.

Irmscher, H. D. (1977), "Johann Gottfried Herder," in *Deutsche Dichter des 18. Jahrhunderts: Ihr Leben und Werk*, edited by B. von Wiese, Berlin: Erich Schmidt Verlag, 524–550.

Kant, I. (1991), "Idea for a Universal History with a Cosmopolitan Purpose," in *Kant's Political Writings*, 2nd ed., edited by H. S. Reiss, translated by H. B. Nisbet, Cambridge: Cambridge University Press, 41–53.

Kusch, M. (2016), "Wittgenstein's *On Certainty* and Relativism," in *Analytic and Continental Philosophy: Methods and Perspectives. Proceedings of the 37th International Wittgenstein Symposium*, edited by H. A. Wiltsche and S. Rinofner-Kreidl, Berlin: De Gruyter, 29–46.

Meinecke, F. (1972), *Historism: The Rise of the New Historical Outlook*, translated by J. E. Anderson and rev. H. D. Schmidt, London: Routledge and Kegan Paul.

Menze, E. (2002), "Herder and Prejudice: Insights and Ambiguities," *Herder Jahrbuch / Herder Yearbook* 6, edited by K. Menges and R. Otto, Stuttgart and Weimar: Verlag J. B. Metzler, 83–96.

Rachels, J. (2007), *The Elements of Moral Philosophy*, 5th ed., Boston: McGraw Hill.

Sikka, S. (2011), *Herder on Humanity and Cultural Difference: Enlightened Relativism*, Cambridge: Cambridge University Press.

Spencer, V. A. (1998), "Beyond Either/Or: The Pluralist Alternative in Herder's Thought," *Herder Jahrbuch / Herder Yearbook* 4, edited by H. Adler and W. Koepke with S. B. Knoll, Stuttgart and Weimar: Verlag J. B. Metzler, 53–70.

—— (2007), "In Defense of Herder on Cultural Diversity and Interaction," *The Review of Politics* 69: 79–105.

—— (2012), *Herder's Political Thought: A Study of Language, Culture and Community*, University of Toronto Press, Toronto.

Taylor, C. (1992), "The Politics of Recognition," in *Multiculturalism and "The Politics of Recognition": An Essay*, edited by A. Gutmann, Princeton: Princeton University Press, 25–73.

Walzer, M. (1994), *Thick and Thin: Moral Arguments at Home and Abroad*, Notre Dame, Ind.: University of Notre Dame Press.

Williams, B. (1972), *Morality: An Introduction to Ethics*, New York: Harper and Row.

—— (1985), *Ethics and the Limits to Philosophy*, Cambridge, Mass.: Harvard University Press.

Williams, R. (1983), *Keywords: A Vocabulary of Culture and Society*, London: Fontana Press.

Zuckert, R. (2015), "Adaptive Naturalism in Herder's Aesthetics: An Interpretation of 'Shakespeare'," *Graduate Faculty Philosophy Journal* 36: 269–293.

13 Socializing knowledge and historicizing society

Marx and Engels and the manuscripts of 1845–1846

Terrell Carver

Introduction

Marx's now-famous manuscript "Theses on Feuerbach" of early 1845 record—apparently as "notes to self," rather than for publication—an extraordinary attack on idealisms and materialisms as understood to date. His declarations there, and in subsequent rough manuscripts of 1845–1846, of a "new" post-Feuerbachian view of philosophy and politics have led some commentators to conclude that in those manuscript works, he broke free from subject-object epistemologies, whether founded on the perception of objects by individual minds or on "mind" that is supervenient to individuals and objects alike. In Theses 1, 9, and 10, Marx dismisses "all previous materialism" as "contemplative" and opts instead for a "new [one]." But in Thesis 1, he also briefly praises "idealism" for having "set forth" the "active side," i.e., "sensuous human activity, practice," albeit "abstractly" (1976 [written early 1845], 3–5).

A particular Wittgensteinian strand of commentary on Marx resolves the apparent contradiction in critiquing both materialism and idealism, and the apparent oddity of remaining in some sense a materialist. Ted Schatzki undertakes a careful parallel interrogation of Marx's early texts and Wittgenstein's later ones in order to draw out two commonalities: activity as "the central feature of human life," as opposed to "contemporaneous forms of idealism" that located mind and meaning elsewhere (2002, 56); and activity as "inherently social," thus linking material production and physical consumption with language and culture (2002, 58). Summing up, Schatzki argues that for Marx, "productive action is social because it occurs as part of, and its intelligibility rests on, understandings carried in a collective complex of activity." And for Wittgenstein, "actions are practices, usages, institutions," making action "inherently social" rather than quintessentially individual (2002, 59). Moreover, for both thinkers, these conceptions are rooted in a "natural history," since neither "brooks an opposition between nature and society": society "is part of human nature," and "nature is part of human society" (2002, 60). On this view, Marx, in his "Theses on Feuerbach," conceptualizes knowledge-making as practical social activity, thus removing subject-object epistemologies from his "new" materialism and substituting a historical and developmental "active side" for static perception.

This chapter draws out the textual basis for this novel view by analyzing passages—in an entirely new way—from the now canonical but actually factitious "book," *The German Ideology*. It then explores the implications of Marx and Engels's theory of knowledge for the problem of relativism.

Politico-philosophical manuscripts of 1845–1846

Within the critiques of German idealists and would-be socialists that represented Marx's and Engels's publishing projects of mid-1845 to mid-1846, only very few items actually reached the press, leaving a large number of handwritten pages famously abandoned (Carver and Blank 2014a). The very roughest of these were, quite surprisingly, fabricated by David Riazanov into a "Feuerbach chapter" in 1923–1924 in a bid to gain political support and financial backing for his projected *Gesamtausgabe* (complete edition) of the works of Marx and Engels, edited and published in the original languages to the highest standards of bibliographical science (Carver 2010). Following on from one of the first volumes to be published in that monumental project, successive editions of the so-called book *The German Ideology*, supposedly by Marx and Engels as sole joint authors, and opening with a "chapter" called "1. Feuerbach," have, from 1932, commenced with this material, albeit variously rearranged and re-edited over the years. Following Riazanov's novel claims for what had formerly been dismissed as inconsequential detritus, the "Feuerbach chapter" has been widely regarded ever since as of the highest interest in relation to the "materialist interpretation of history."[1] This phrase was Engels's locution of 1859, rather than Marx's own summary phraseology (Carver 2003, chapters 5–7; 2010, 109).

However, editorial practice requiring a "smooth text of the last hand" has largely discouraged and indeed effectively prohibited the use of these very rough extractions from draft *ad hominem* critiques to pursue the way that Marx and Engels, working as individuals, developed the critical insights of the "Theses on Feuerbach" in an innovative manner (Carver 2010).[2] By following their authorial drafting in detail from the manuscript pages, however, we can see that the two were developing the abstract ideas occurring in the "Theses on Feuerbach" into the vocabulary of a new *Ansicht* (outlook) or *Auffassung* (conception) (Carver 2015). In their draft thoughts, the two used this new understanding to reconceptualize the conventional terms through which ontology and epistemology are understood, albeit in their preferred mode of political confrontation.

Because this manuscript material is so rough (and uniquely so in the Marx–Engels *Nachlass*), it is possible to trace a movement in the two authors' vocabulary from tropes of empirical reference, using a classic subject–object epistemology, to terms that render a novel theory of knowledge that proceeds on a new basis entirely. This new theory transcends both idealism and materialism, conceived as epistemologies proceeding from presumptions of consciousness and its objects, whether argued through in a Lockean, Kantian, or Hegelian manner, as outlined above.

In this chapter, I use an English-language rendition of selections from these very rough manuscript drafts, recently transcribed from German and presented in a unique way in my own English translations, to illustrate and expound a nascent view of knowledge as social practice, practice as politics, and politics as future-making. Epistemology is thus socialized and historicized "all the way down": on this view, knowledge, and therefore the criteria for truth or falsity, can no longer be construed in any way that abstracts from human activities located sequentially in time.

Thinking rather than "thought"

Conventional reception and bibliographical methods presume the doctrinal character of Marx and Engels's *thought*, and very usually "Marx and Engels's" *joint* thought (see Carver 1984), and nearly all critical scrutiny has been directed hermeneutically toward that goal. However, transcribed here as a "variant-rich text," readers can instead see how Marx and Engels were *thinking*. This approach thus sidelines any rush to determine the final or "last hand" content of their *thought*—which is the usual political and scholarly goal—at this point or later. A focus on "thought" usually spurs most commentators to summary accounts of their views and a characterization of these thinkers as doctrinaires.[3] By contrast, I focus here on the *thinking*, which allows the authors to range widely over any number of questions in terms of their own priorities, given the coded philosophical nature of political encounters current at the time.

In that way, this chapter starts afresh, leaving the "materialist interpretation of history" to look after itself, and refusing to debate the binary line that supposedly divides science from philosophy, given that both these intellectual *idées fixes* are anachronistic to the period. Rather, the working hypothesis here is that the two-handed manuscript drafts available in German transcription today can be utilized—*sans* the presumptive "smooth text of the last hand"—to display a drift in the authors' vocabulary away from the classical dichotomies through which epistemology and ontology (whether materialist or idealist) had formerly been understood, and through which most philosophical discussions still operate.

The analysis presented here demonstrates that the questions the two were asking of themselves, and attempting in tentative, exploratory, and dialogical fashion to answer, are still crucially problematic in philosophy, as well as in politics. These concern the nature of humanity, the character of social activity, what makes generalizations plausible, truthful, and persuasive, and thus classic questions in epistemology and ontology. As Marx and Engels approach these questions, they argue a polemical transcendence of the traditional dichotomies and antinomies. Their "outlook" conceives of the meaningful world as thoroughly humanized and historicized, rather than as a timeless and indeed usually abstract conjunction of mind with matter.

However, we need to bear in mind that these fragments derive from polemical works in which Marx and Engels were excoriating philosophical idealists, who were a *political* presence at the time, so their adoption of "materialist" as a moniker was a hostile declaration rather than some obvious philosophical move in agreed *academic* territory. I argue here that these rough manuscript fragments are thus interesting to the extent that they explicate Marx's critique—as rendered in the notes-to-self "Theses on Feuerbach" of the immediately preceding months—of previous materialisms. This development occurs in a discursive style, albeit in a two-handed drafting process, in what Riazanov termed the "main manuscript" (Carver 2010, 120). The messy drafts thus work through the famous apothegms that were indeed directed at Feuerbach specifically in Marx's "Theses," though in the materials considered here Feuerbach's name occurs only as an adjunct to critical barbs directed at other—now less well-known—targets. The variant-rich presentation of the so-called main manuscript sheets tells us in a number of places something about how this "new" materialism—among other possible considerations, of course—was put into words, word-by-corrected-word.

Presentation and analysis[4]

Starting with the two sheets that remain of what is presumed to be *Bogen* 1[5] of an otherwise lost, early critique of Bruno Bauer—and thus the opening fragment of the first part of what became known as the "main manuscript" of the "Chapter" "I. Feuerbach"[6]—here is Engels's hand in full polemical flow:[7]

> Naturally of course we will not take the trouble to enlighten t̶o̶ our wise philosophers with the fact that the "liberation" of "man" does not get a single step further when they have dissolved philosophy, theology, substance & all that foolery into "self-consciousness," when they have liberated "man" from domination by these phrases to which he had never been in thrall; that it is not possible to achieve actual liberation other than in the actual world & with actual means …
>
> (Carver 2015, 711–712)[8]

And here is Marx's summary reduction—making the same point without the rhetorical turns of sarcastic polemic ("we will not trouble to enlighten …") that are present in the other, left-hand column in Engels's hand. Feuerbach is addressed throughout as superior to "his rivals" in Young Hegelian circles (e.g., Bruno and Edgar Bauer), but still suffering from the same affliction, namely being a philosopher, protestations—and in Feuerbach's case, a self-declared "materialism"—to the contrary notwithstanding.

220   *Terrell Carver*

> Just like his rivals Feuerbach believes{himself} to have transcended phil-
> osophy! ~~The act{ual}~~ The struggle against general conceptions, which have
> previously oppressed the individual, summarizes the standpoint of German
> philosophical criticism. We maintain that this struggle, pursued in this
> manner, is itself founded on philosophical illusions of the sovereignty of
> general conceptions.
> Philosophical and actual liberation.
> Man. Individuality. The Individual.
> Geological, hydrographical etc. conditions.
> The human body. Needs and labor.
>
> (Carver 2015, 712–713)[9]

Possibly, Marx's deletion—"~~The act{ual}~~"—allowed him to finish the pejorative characterization of idealism and idealists as a matter of "**struggle**" and
then proceed to a clear contrast: "**Philosophical and actual liberation.**" And
after that, he advances—interestingly—to conditions conventionally understood as material and thus as the "external" setting for humans ("**Geological,
hydrographical etc. conditions**"), then he moves to humans "materially"
considered ("**body**") and finally onward to the social yet still "material" constitution of individuals and societies ("**Needs and labor**"). This sets out an
argumentative trajectory from the speculative realm of (merely philosophical) ideas to a realm of actuality, where materiality and sociality intersect in
practical (rather than abstract) ways, and where an abstraction such as "man"
does not figure as a mystifying syntactical agent, but rather more flesh-and-
blood conceptualizations take his [*sic*] place.

Actuality is thus "filled in" by Marx, not so much with materiality and
therefore "objects" as with the sociality and history of human beings, understood quite apart from the conventionally presumed philosophical parsing
of reality into dichotomous categories of matter/mind or objects/ideas. In
a sense, the left-hand column draft in Engels's hand—"actual world & with
actual means"—has been resolved into an agenda for specifics in Marx's right-
hand column commentary. Here, we have an opportunity to follow the actual
vocabulary through which this philosophical (and also anti-philosophical)
move is pursued.

Engels's left-hand draft indeed becomes much more specific in terms of
Marx's "**Needs and labor**" by historicizing the industrializing societies with
which he was personally quite familiar:

> that it is not possible to achieve actual liberation other than in the
> actual world & with actual means, that slavery cannot be transformed
> {*aufheben*} without the steam-engine & spinning machines, serfdom
> without improved agriculture, that in any case men cannot be liberated
> so long as they are not in a position to obtain food & drink, shelter &
> clothing ~~adequate~~ sufficient in quality & quantity.
>
> (Carver 2015, 713–714)[10]

While this view of human history is not completely novel, and indeed reflects the historical mode through which political economy had been developing since the seventeenth century, the communists' twist here is, of course, the view that the future offers "liberation." This is spelled out through the "new" materialism of historically sequential practical meaning-making, rather than through conceptual binaries of ontology/epistemology, materialism/idealism:

> "Liberation" is a historical action, not a conceptual action, & it is accomplished through historical relations, through the state of industry, of trade, of agriculture, of social interaction {relation}s{,} ...
>
> (Carver 2015, 714)[11]

Using history to project a future is a familiar move with considerable rhetorical value, and indeed in that way history is never simply of the past. The present is the past as far as it has got, and the future is where the present is going (or should be going if moved along politically as Marx and Engels were intending). History, of course, could contrarily in idealist terms be a history of concepts, or a chronologically arranged account of conceptual transformations, which was indeed the accusation that Marx and Engels were levelling against their opponents, in this case, "Saint Bruno" Bauer.

Marx and Engels were evidently working to distinguish themselves from such idealists, not by declaring a conventionally oppositional subject-object materialism in ontological and epistemological terms, but by defining actuality as the developmental trajectory (past, present, future) of human productive activities, technologies, industries. These are of course words and in some sense abstractions. Where, then, is the epistemological dividing line? What is it about these words that references actuality more truthfully than idealist assertions (as Marx and Engels portray them) of transformations involving "man," "self-consciousness," "substance," and "liberation," or indeed more truthfully than objectively referential and materially "grounded" empiricisms?

Reading through the next sequence of sheets, as we have them, *Bogen* 6 to *Bogen* 11, it becomes striking that the traditional interpretive strategy suggested by Engels (1886, 520), and later endorsed by Marx's first biographer, Franz Mehring (1918), namely, that mere polemic can be excised from substantive (and in their terms, validly philosophical) content, can usefully be reversed. This is not to say that the specific points made by Marx and Engels against Bauer (and against Bauer's version of Feuerbach, and independent of that, against Feuerbach in Marx and Engels's understanding) are of particular interest in themselves, but rather that Marx and Engels's substantive theses on humanity, history, modernity, and a communist future develop in these fragments as political but still (anti)philosophical points through and through. This is by way of contrast to truths that would make sense whatever the context, or indeed require no context, such as are conventionally privileged in philosophy. What emerges in the manuscript discussions is that the two have a common position—which they are working out in some sense jointly—albeit one conceived pre-eminently as a political and (anti)philo-sophical "outlook" or "conception."

Indeed, the tenor of the argumentation is such that extraction of Marx and Engels's views as "theory" (whether a philosophical one or a theory of history) would be a regression to the very position—excoriated as both ideological and typically German—that they were at such pains to attack in their renewed critique of the "critical critics."[12] The nub of the matter was not so much that these idealist (and therefore ineffectually critical) philosophers were thinking the wrong things because they were thinking the wrong way, but that they were doing politics the wrong way—hence thinking the wrong way—and thus merely encouraging others to be just as wrong-headed and—so Marx and Engels were arguing—ineffectual in every sense.

Interesting and provocative ideas certainly can be extracted from these fragments of polemic, though there is a contextual and interpretive dis-junction in doing so. Marx and Engels (in this period) can be made into methodologists of philosophy and/or history (or indeed into many other things), but doing so traps the commentator in the very critique that the two are mounting. This was a relentless critique of political posing and posturing, of self-deceiving fantasies of potency belied by evident social realities and predictable economic developments, namely, the spread of industrialized commodity production, global markets, and relentless commercialization. Of course, there could be spirited defenses today of "Bauer & Co.," as well as cynical judgements that Marx and Engels were never on to a winner with their own political strategy (and were thus self-deluding in their own ways, which were not entirely dissimilar, given the difficulties in pursuing any kind of socialist/communist activisms at time). However, for the present exercise—exploring their (anti)philosophy as a novel transcendence of subject/object epistemologies and ontologies—we can let the pair have it their own way on their own manuscript pages.

Here in the passage below, we have what might be an interesting illustration of a move from stating what is the case in very general and abstract terms—as a philosopher might do—to a more direct approach (via a strike-through), namely stating what a "practical materialist" should be *doing* politically. In this way, Marx and Engels argue a circular consistency—against "contem-plative" materialisms—that meaning-making comes through activity, and activity is meaning-making. Note the force of Marx's emphasis:

> in reality {it} ~~is a matter of~~ & for the **practical** materialists, i.e. the **communists**,[13] it is a matter of revolutionizing the existing world …
>
> (Carver 2015, 715–716)[14]

And note below that in his insertions, Marx is devaluing the language of the-orizing with scare quotes, and excising an abstraction in favor of a concrete reference:

Feuerbach's ~~theoretical conception~~ "conception" of ~~perceptibility~~ the perceptible world is limited on the one hand to merely viewing it, & on the
other {insertion} to merely {end insertion} feeling {it} …

(Carver 2015, 716)[15]

Finally, here is Marx's insertion (again, "he" is Feuerbach) nailing down this
contrast—between a philosopher's very general abstractions and a "concrete"
reference to a politically potent alternative:

> {he} considers "<u>man</u>" instead of "actual historical man". "<u>Man</u>" is in
> reality "German man".

(Carver 2015, 716)[16]

In these passages below Marx and Engels are criticizing Feuerbach for merely
hinting at what they themselves are *stating* directly—as opposed to more egre-
gious ideologists who have not advanced even to the point that Feuerbach had
reached with his hints. Feuerbach is thus rather a side-issue, i.e., a stick to beat
Bauer with.

The critique from Marx and Engels throughout is two-fisted: German
ideologists have the wrong philosophy and wrong practice (and indeed, they
are philosophers, so the academically vocational and non-practical outlooks
go together), and they have the wrong history (because philosophy in general
is abstractly timeless, so they really have no historical sense at all, hints and
protestations notwithstanding). In the lines below, "He" is again Feuerbach:

> He does not see how the perceptible world surrounding him is not a thing
> handed down directly from eternity, staying always the same, but rather
> the product of industry & of social conditions & to be sure in the sense
> that it is a historical product, the result of the activity of a whole series
> of generations, …

(Carver 2015, 716)[17]

From this Marx/Engels perspective, there is little point in a materialism
of "things" or objects of perception as such, given that things/objects are
asserted to have human *histories* of production, thus making the argument
not so much about what things are but rather how practical activities in
society make things what they are, i.e., what they are known *to be*. Effectively,
this view dissolves and transcends the most entrenched dichotomy of post-
seventeenth-century philosophy, that of matter/consciousness as exclusionary
categories, one (consciousness) "knowing" the other (matter) as already what
it *is*, such that knowledge of it comes through perception.

Perhaps the Marx-Engels position could be characterized as an epistemology of action, rather than of knowledge as such, i.e., truth is an effect of
meanings made in practice. The "NB" in Engels's hand in the right-hand
column quoted below is rather wordy and repetitious, but gets to the point in

the end, implying that "spectacles," a metaphor for philosophizing, produce not just impaired vision but failure to make sense of human experience in a more politically progressive manner than merely "spectating," as an idealist and "contemplative" materialist would:

> NB. F[euerbach's] mistake is not that he subordinates the immediately apparent, <u>perception</u>, to the perceptible actuality attested by precise investigation of perceptible circumstances, but that he cannot in the end [*in letzter Instanz*] cope with perceptibility except by considering it with the "eyes," i.e., through the "spectacles," of the <u>philosophers</u>.
>
> (Carver 2015, 717)[18]

The following lengthy passage produces a clear, if philosophically still controversial, definition of what constitutes an empirical fact, which Marx and Engels argue is not a reference to what something is but rather to "what has happened" in terms of practical, social activity in order to produce it historically. This means that an empirical fact is not a linguistic representation of, and thus a conceptual reflection of, an object that is discretely given to perception and simply is what it is in itself. Marx's insertions ram home the point polemically, saying that Bauer & Co. have not grasped the essentially and profoundly historical character of things/objects, even supposedly natural ones, because they presume an antithesis between (timeless) nature and (happenstance) history:

> Moreover in this conception ~~also the~~ of things as they actually are & have happened, every profound philosophical problem resolves itself quite simply into an empirical fact, as is shown even more clearly below. E.g. the important question of the relation of man to nature ~~on which~~, {*insertion*} (or especially ~~the "relation between~~ as Bruno says (p. 110)[19] the "antitheses in nature and history," as if {*further insertion*} the two were quite separate "things," {*end further insertion*} {as if} man is not always confronted with a historical nature and a natural history,) {*end insertion*}
>
> (Carver 2015, 717)[20]

The argument above is that for humans, their world is "**a historical nature and a natural history**." Marx's chiasmus is worth pondering, precisely because to be human is to be historical "all the way down" and thus to make a history of material/natural objects (not just a perception) and to see material/natural objects within a historical perspective (not a timeless one). Rounding off this revolution, the two authors in the passage below collapse a quoted phrase linking the two abstractions man and nature—the phrase has a philosophical, even quasi-religious, ring to it—into an utterly mundane and everyday concept of industry, which has its ups and downs historically:

the much ~~famed~~ vaunted "unity of man with nature" has always existed in industry & has existed variously in every epoch depending on the lesser or greater development of industry, ...

(Carver 2015, 718)[21]

The question of the temporal priority of "nature" over "man" arises below within a long insertion in Engels's hand, as it would in a philosophical argument that "man" and "nature" must be considered distinct simply because the former arose prior to, and independently of, the latter. But then this conventional treatment is brusquely replaced with a different understanding, one that locates meaning in the current political setting, rather than in some timeless elsewhere:

> ~~For~~ In any case the {temporal} priority of {*insertion*} external {*end insertion*} nature remains intact here, ~~& it is no accident for us &~~ & in any case ~~this nature no distinction~~ all this has no application to the first men produced through spontaneous generation; this distinction, however, only has meaning in so far as one considers man to be distinguished from nature. Moreover this nature, which precedes human history, is really not ~~Feuerbach's, in which~~ the nature in which Feuerbach lives, not the nature which no longer exists anywhere today except perhaps ~~in the interior of newly f{ormed}~~ on isolated Australian coral islands of recent origin ...
>
> (Carver 2015, 718)[22]

Another passage in Engels's hand much later in this set of fragments—and possibly written somewhat later in the compositional processes anyway—puts this issue with great clarity. Note also the strikethrough on "proof," which might have philosophical connotations of certainty via abstract reasoning. This move further devalues Feuerbach's method, which—according to Marx and Engels—was to universalize abstractly from (ever-shrinking) examples of supposed certainties founded on their timelessness, and thus in contradistinction to the visible realities of human history and the historicizing vision that the two were developing by contrast:

> Feuerbach therefore never speaks of the human world but rather he flees every time into external nature, & to be sure into the nature which has not yet been brought under human control. But with each new invention, each advance of industry a new patch is detached from this terrain, & the soil, from which grow the ~~proof~~ [*Beweis*] examples for similar Feuerbachian propositions, {hence the terrain} is thus becoming ever smaller.
>
> (Carver 2015, 718–719)[23]

The historicity of the human–nature relationship could not be clearer than the way it is put in the passage above. What is "external" is merely *not yet* under human control and thus emphatically not a realm of timeless certainty on which human reasoning could rely, or could indeed be "grounded." One of Marx's pithy insertions in the passage below also puts the conclusion unmistakably and in simple terms:

> Feuerbach has in any case a big advantage over the "pure" materialists because he ~~also realizes how~~ realizes how man too is "a perceptible object"; however, {*insertion*}[24] **apart from the fact that he only conceives of him as "~~perceptible~~ a "perceptible object" not as "perceptible activity"** ...
>
> (Carver 2015, 719)[25]

Or, in other words, any conception of humanity must start by conceiving of humans, not as objects in any discrete or "material" sense, but as always already immersed in activities, which are themselves necessarily historical and thus time- and sequence-dependent.

In the passage below in Engels's hand, the critique of Feuerbach's ultimate "philosophism" (which in this context is, of course, a way of trouncing Bauer & Co. as even worse than Feuerbach) is attacked in another of Marx's insertions, where he moves from the pithy to the earthy, and in a potentially even more revolutionary way. Feuerbach, it says below in Engels's hand,

> only gets as far as recognizing the "actual, individual, embodied men" in terms of emotion, i.e. he ~~arrives at~~ knows no other "human relations" "of man to man" other than love & friendship, {*insertion*} **and idealized at that. There is no critique of present-day loving relations.** {*end insertion*}
>
> (Carver 2015, 719)[26]

Passages below in Engels's hand struggle mightily to get from the material—but also language-based—character of human activities, and on to an explanation of consciousness as such. This raises the mind/matter dichotomy and leads to an unhelpful excursus on the materiality of language itself: it is said to comprise three aspects or moments, subsequently altered to four without any clear amendment:

> Only now, after we have considered four moments, four aspects of original, historical relations do we find that man ~~among other things also has "mind", & that this "mind" "manifests" itself as "consciousness"~~ {*insertion*} **also has "consciousness"** {*end insertion*}. But even this {is} not from the outset "pure" consciousness. The "mind" has from the start the curse of being "burdened" with matter, which ~~here in the form of vibrating layers of air, sounds, in short, language~~ occurs here in the form of vibrating layers of air, sounds, in short, language. Language is as old as consciousness—language is practical, actual consciousness existing for other men as well {*insertion*}, only therefore does it also exist for me myself {*end insertion*}, ...
>
> (Carver 2015, 721)[27]

A summary comment from Marx makes short work of this apparent problem: humans-in-activity simply *are* a union of physicality and consciousness:

> **Men have history, because they must <u>produce</u> their life, and indeed must do so in a <u>specific</u> way; they ha{ve} this is given by their physical organization; just the same as their consciousness.**
>
> (Carver 2015, 721)[28]

Here is an anti-philosophical manifesto from the latter part of the "main manuscript" text, dissolving epistemology into history:

> {*insertion*} The distinction between {*end insertion*} what is personal to the individual & what is contingent to the individual is not a conceptual distinction but rather a historical fact. This distinction has a different significance at different times, e.g. the medieval estate as something contingent to the individual in the 18th century, also the family, more or less. It is not a distinction that we have to make for each era but rather each era makes the distinction itself out of the different elements that it finds to hand, & to be sure not according to a concept but rather forced by the material interactions of life.
>
> (Carver 2015, 721)[29]

Note the interesting correction below of "things" to "practice," i.e., moving the language from a discourse of material object to one of human activity:

> The division of labor only becomes an actual division at the moment when a division of ~~mental & material~~ material & mental labor takes place. From this moment onwards consciousness <u>is able</u> to conceive of itself as something other than the consciousness of existing ~~things~~ practice, ~~something actual~~ actually representing something without representing an actual thing—from that moment onwards consciousness is in a position to emancipate itself from the world & to ascend to ~~pure~~ the formation of "pure" theories, theology{,} philosophy{,} {*insertion*} morals {*end insertion*} &c.
>
> (Carver 2015, 722)[30]

The Marx/Engels "conception" merges epistemology with ontology, what is with how we know. In the passage below, the insertion by Marx perhaps explains what "material" is otherwise ambiguously indicating:

> at each stage there is to hand a material result, a sum of productive forces, {*insertion*} **a historically created relation to nature and of individuals to one another** {*end insertion*}, ...
>
> (Carver 2015, 723)[31]

Thus, "a material result" is not a sum of things (even including human bodies) but rather a complex socio-material *formation* supervenient to any supposed parsing of reality into nature/matter/things and mind/ideas/consciousness.

Marx's drafting is possibly more consistent than Engels's in avoiding ambiguous references to "material" and sticking with the praxis terminology:

> **The "conception," the "representation" of these specific men concerning their actual praxis is transformed into** ~~the actual defining and active essence~~ **the sole defining and active power** which controls & defines the praxis of these men.
>
> (Carver 2015, 723)[32]

Notice above the strikethrough on "essence," which is a classically philosophical term, and Marx's rephrasing in terms of power, which is much more this-worldly and socially experiential.

The problem of relativism

Marx and Engels deserve considerable credit, not just for developing an *"Ansicht"* (outlook) or *"Auffassung"* (conception) on humanity, civilization, industry, politics, society, change, and the future, but for exposing the kinds of questions that need to be asked in order to get that discussion going in a productive way. They also expose the kinds of supposedly clarificatory distinctions that merely get in the way, philosophical traditions and authorities notwithstanding. How, then, does an ontology/epistemology resting entirely on practical, social activity sequential in history, rather than on ontological "foundations" in matter or mind/*Geist*, relate to the problem of relativism?

The problem of relativism arises from a presumption that philosophers must work from a basis of ontological certainty about what "is," and epistemological certainty about how to "know" it. Knowledge is thus guaranteed truthful because it arises within this truth-making apparatus which is anchored in certainty. It follows that relativism, or relativist views, *depart* from this apparatus—which is common to both materialisms and idealisms—precisely because they offer no basis in certainty from which knowledge for well-founded judgements can arise. Anyone subscribing to such relativisms thus has no basis in certainty from which to argue judgments, so it follows that all such judgments are merely relative to each other rather than absolute, because certain. The problem of skepticism arises with respect to how exactly a basis in certainty is established, so skepticism is distinct from relativism, which—one way or another—denies that such a basis can be established at all. Relativism and skepticism are thus issues that derive from presumptions that knowledge must be derived from, and so founded on, certainty.[33]

Thus, in order to understand the (anti)philosophy offered by Marx and Engels one needs to transcend the terms of the problem. The problem of relativism only arises from a presumption that knowledge—and therefore truth—must have foundations outside of practical, social activity sequential in history, whether these foundations are "grounded" in matter (in a commonsensical or scientific sense) or ideas (whether as individualized consciousness or as pantheistic *Geist*). Given the paradox that foundations cannot be "known" independently of knowledge, and that the world cannot already be imbued with human or human-accessible meanings, unless imparted there by a human-like Creator (Rorty 1989, 4–7), it follows that the problem of relativism is at a distinct and depoliticizing distance from the practical, social activity, sequential in history, that is privileged by Marx and Engels over philosophical concerns. For them, these concerns displace the exigencies of political analysis and action, and worse still, constitute themselves discursively—as in the person and writings of "Saint Bruno" Bauer—as a pathetic masquerade. Marx's Thesis 2 "on Feuerbach" states this (anti-) philosophical transcendence succinctly:

> The question whether objective truth can be attributed to human thinking is not a question of theory but is a *practical* question. Man must prove the truth, i.e., the reality and power, the this-worldliness of his thinking in practice. The dispute over the reality or non-reality of thinking which is isolated from practice is a purely *scholastic* question
>
> (Marx 1976 [written 1845], 3; emphasis in original).

The analysis of Marx and Engels's subsequently composed "main manuscript" drafts, undertaken above, shows how they worked this insight into the polemical arguments that constituted their activist interventions into the politics of the time. Transcending the epistemological framing of conventional ontologies—both materialisms and idealisms—and indeed formulating an anti-philosophy, as they did, rendered the problem of relativism—from the perspective of their "outlook" or "conception"—redundant.[34]

Notes

1 The latest rendition of these somewhat miscellaneous works and manuscripts of 1845–1846, published in *Marx-Engels-Gesamtausgabe* (2017), reproduces quite faithfully Riazanov's editorial fabrication in terms of ordering of items and assumptions about their significance.

2 Note that Engels (1886, 520) seems to report a discovery of the "Theses on Feuerbach" rather than a rediscovery, suggesting that he was unaware of the notebook pages when he was working with Marx on the manuscript polemics.

3 In their quite different ways, these standard works are very typical of the approach: Berlin 2013 [1939]; Althusser et al. (1965); McLellan (1973); Cohen (1978). In more recent biographies this approach is disavowed but the results of it as undertaken by others are then uncritically recounted, sometimes in garbled fashion; for example, Sperber (2013), Stedman Jones (2016).

4 This section adapts material previously published in Carver and Blank (2014b, 4–29), and Carver (2015, 710–724), used with permission.

5 *Bogen* are large printer's sheets in a four-page "greeting-card" format.
6 Though given editorial proclivities to collage, this fragment was published in different locations in the various "editions;" for a detailed treatment of the issues involved, see Carver and Blank (2014a).
7 For the passages presented below, here is a brief *apparatus criticus*:

~~deletion~~ = excised word or phrase
{*insertion*}/{*end insertion*} = insertions by Marx or Engels
[square brackets] = insertions by the editors of *Jahrbuch 2003*
{braces} = editorial insertions by TC
roman typeface = Engels's handwriting
bold typeface = Marx's handwriting
<u>underline</u> = emphasis in the manuscript

Punctuation and capitalization are in conformity with English usage, but I have taken the transcription of the German manuscript into account as much as possible.

8 01 *Bogen*, 01 *Seite*, Left Column; *Jahrbuch 2003*, *Text*, 6; Carver and Blank 2014b, 4–5, 34–35. Extracts are identified by *Bogen* number (Engels) and *Seite* number (Marx) where extant, and by column location (Left and Right), together with references to the published German transcription in the *Text* volume of *Jahrbuch 2003*.
9 01 *Bogen*, 01 *Seite*, Right Column; *Jahrbuch 2003*, *Text*, 6; Carver and Blank 2014b, 5, 34–35.
10 01 *Bogen*, 01 *Seite*, Left Column; *Jahrbuch 2003*, *Text*, 6; Carver and Blank 2014b, 6, 36–37.
11 01 *Bogen*, 01 *Seite*, Left Column; *Jahrbuch 2003*, *Text*, 6; Carver and Blank 2014b, 6, 36–37.
12 The "critical critics" (i.e., "Bruno Bauer and Company") were the butt of Engels and Marx's (1845, 3) political satires in *The Holy Family: Critique of Critical Criticism*.
13 Marx's emphasis at these two points; *Jahrbuch* 2003, *Apparat*, 213, ref. 7.17–18 l.
14 06 *Bogen*, 08 *Seite*, Left Column; *Jahrbuch 2003*, *Text*, 7; Carver and Blank 2014b, 7–8, 44–45.
15 06 *Bogen*, 08 *Seite*, Left Column; *Jahrbuch 2003*, *Text*, 7; Carver and Blank 2014b, 8, 44–45.
16 06 *Bogen*, 08 *Seite*, Left Column; *Jahrbuch 2003*, *Text*, 7; Carver and Blank 2014b, 8, 44–45.
17 06 *Bogen*, 08 *Seite*, Left Column; *Jahrbuch 2003*, *Text*, 8; Carver and Blank 2014b, 8, 46–47.
18 06 *Bogen*, 08 *Seite*, Right Column; *Jahrbuch 2003*, *Text*, 8; Carver and Blank 2014b, 9, 46–47.
19 Bauer (1845).
20 08 *Bogen*, 09 *Seite*, Left Column; *Jahrbuch 2003*, *Text*, 9; Carver and Blank 2014b, 10, 50–51.
21 08 *Bogen*, 09 *Seite*, Left Column; *Jahrbuch 2003*, *Text*, 9; Carver and Blank 2014b, 10, 50–51.
22 08 *Bogen*, 09 *Seite*, Left Column–08 *Bogen*, 10 *Seite*, Left Column; *Jahrbuch 2003*, *Text*, 10; Carver and Blank 2014b, 10–11, 54–57.

23 11 *Bogen*, 29 *Seite*, Right Column; *Jahrbuch 2003*, *Text*, 38; Carver and Blank 2014b, 11, 160–161.

24 The editors of *Jahrbuch* 2003 state that this insertion was written down later than the previous one; *Apparat*, 216, ref. 11.3–6 l.

25 06 *Bogen*, 10 *Seite*, Left Column; *Jahrbuch 2003*, *Text*, 10–11; Carver and Blank 2014b, 11, 56–57.

26 06 *Bogen*, 10 *Seite*, Left Column; *Jahrbuch 2003*, *Text*, 11; Carver and Blank 2014b, 12, 58–59.

27 07 *Bogen*, 13 *Seite*, Left Column–07 *Bogen*, 14 *Seite*, Left Column; *Jahrbuch 2003*, *Text*, 15–16; Carver and Blank 2014b, 13–14, 72–75.

28 07 *Bogen*, 13 *Seite*, Right Column; *Jahrbuch 2003*, *Text*, 15–16; Carver and Blank 2014b, 14, 72–73.

29 89 *Bogen*, 60 *Seite*, Left Column; *Jahrbuch 2003*, *Text*, 79–80; Carver and Blank 2014b, 14–15, 324–325.

30 07 *Bogen*, 15 *Seite*, Left Column; *Jahrbuch 2003*, *Text*, 17; Carver and Blank 2014b, 15, 78–79.

31 10 *Bogen*, 24 *Seite*, Left Column; *Jahrbuch 2003*, *Text*, 30; Carver and Blank 2014b, 20, 132–133.

32 10 *Bogen*, 25 *Seite*, L Column; *Jahrbuch, 2003*, *Text*, 32; Carver and Blank 2014b, 20, 138–139.

33 Wittgenstein (1969) is, of course, a meditation on this theme.

34 For a lengthy study that considers Marx's development as an intellectual and argues contrary to the above discussion that he ultimately failed to "leave philosophy," see Brudney (1998).

References

Althusser, L., É. Balibar, R. Establet, P. Macherey, and J. Rancière (1965), *Reading Capital*, translated by B. Brewster and D. Fernbach, London, Verso, 2016.

Bauer, B. (1845), "Charakteristik Ludwig Feuerbach," *Wigand's Vierteljahrsschrift* 3: 86–146.

Berlin, I. (1939), *Karl Marx: His Life and Environment*, 5th ed., Princeton, NJ: Princeton University Press, 2013.

Brudney, D. (1998), *Marx's Attempt to Leave Philosophy*, Cambridge, MA: Harvard University Press.

Carver, T. (1984), *Marx and Engels: The Intellectual Relationship*, Brighton: Wheatsheaf.

—— (2003), *Engels: A Very Short Introduction*. Oxford: Oxford University Press.

—— (2010), "The German Ideology Never Took Place," *History of Political Thought* 31 (1): 107–127.

—— (2015), "'Roughing It:' The 'German ideology' 'main manuscript'," *History of Political Thought* 36 (4): 700–725.

Carver, T. and D. Blank (2014a), *A Political History of the Editions of Marx and Engels's "German ideology manuscripts,"* New York: Palgrave Macmillan.

—— (2014b), *Marx and Engels's "German ideology" manuscripts: Presentation and Analysis of the "Feuerbach chapter,"* New York: Palgrave Macmillan.

Cohen, G. A. (1978), *Karl Marx's Theory of History: A Defence*. Princeton, NJ: Princeton University Press, 2001.

Engels, F. (1886), "[Preface to] *Ludwig Feuerbach and the End of Classical German Philosophy*," in K. Marx and F. Engels, *Collected Works*, vol. 26, London: Lawrence & Wishart, 1990, 519–520.

Engels, F. and K. Marx (1845), *The Holy Family*, in K. Marx and F. Engels, *Collected Works*, vol. 4, London: Lawrence & Wishart, 1975, 3–211.

Jahrbuch (2003), *Karl Marx, Friedrich Engels and Joseph Weydemeyer, Die Deutsche Ideologie: Artikel, Druckvorlagen, Entwürfe, Reinschriftenfragmente und Notizen zu I. Feuerbach und II. Sankt Bruno*, edited by I. Taubert and H. Pelger, issued by the Internationale Marx-Engels-Stiftung, Amsterdam, 2 vols., Berlin: Dietz Verlag.

Marx, K. (1845), "Theses on Feuerbach," in K. Marx and F. Engels, *Collected Works*, vol. 5, London: Lawrence & Wishart, 1976, 3–5.

Marx, K. and F. Engels (2017), *Gesamtausgabe, Werke, Artikel, Entwürfe: Manuskripte und Drucke zur Deutschen Ideologie*, div. 1, vol. 5, edited by U. Pagel, G. Hubmann, and C. Weckwerth, Berlin: de Gryuter.

McLellan, D. (1973), *Karl Marx: His Life and Thought*. London: Macmillan, 1974.

Mehring, F. (1918), *Karl Marx: The Story of His Life*, translated by Edward Fitzgerald, London: John Lane, 1936.

Rorty, R. (1989), *Contingency, Irony and Solidarity*, Cambridge: Cambridge University Press.

Schatzki, T. (2002), "Marx and Wittgenstein as Natural Historians," in *Marx and Wittgenstein: Knowledge, Morality and Politics*, edited by G. Kitching and N. Pleasants, London: Routledge, 49–62.

Sperber, J. (2013), *Karl Marx: A Nineteenth-Century Life*, New York: Liveright.

Stedman Jones, G. (2016), *Karl Marx: Greatness and Illusion*, Princeton, NJ: Princeton University Press.

Wittgenstein, L. (1969), *On Certainty*, edited by G. E. M. Anscombe, Oxford: Blackwell.

14 National Socialism and the problem of relativism

Johannes Steizinger

Introduction

The aim of this chapter is to clarify the meaning and the use of the concept of relativism in the context of National Socialism (NS). This chapter analyzes three aspects of the connection between relativism and NS: The first part examines the critical reproach that NS is a form of relativism. I analyze and criticize the common core of this widespread argument, which is developed in varying contexts, was held in different times, and is still shared by several authors. The second part investigates the ideological debate among Nazi philosophers themselves concerning whether NS is indeed a form of relativism. I focus on the epistemological consequences of Nazi anthropology and analyze both its relativistic tendencies and the strategies used to reject relativism. In contrast to the received view, I argue that Nazi philosophers attempted to overcome both absolutism and relativism. The third part investigates the academic debate on relativism during NS, using the example of the prize question on relativism that was announced by the Prussian Academy of Science in 1936. By examining the academic approaches to the problem of relativism, I also address the question of how broader philosophical debates were related to the core of Nazi ideology. Academic philosophers took the self-understanding of Nazi philosophers seriously. They saw the shared aim of overcoming relativism as an opportunity to collaborate with NS. The brief *conclusion* summarizes the findings of the chapter. I conclude that, in the context of NS, critics, ideologists, and academics understand and use the concept of relativism in the same way.

The received view: National Socialism as relativism

The connection between relativism and NS is often used as a critical argument against both. The weakest form of this argument runs as follows: Anti-relativists claim that relativism is motivated by the conviction that there are many radically different, yet equally valid, epistemic or moral systems. This equal-validity claim ties the relativist to a strong form of tolerance: confronted with a conflicting epistemic or moral system, the relativist has to concede that the other agent is equally justified in her epistemic or moral beliefs. Hence, relativism does not provide us with the normative resources to criticize irrational views such as Nazi racism. We need a normative universalism to confront racist ideologies (Böhler 1988; Tugendhat 2009; Kellerwessel 2014). Such systematic claims are often supported by the historical argument that the relativism of post-Hegelian philosophy indeed paralyzed the moral consciousness of German intellectuals during Weimar Republic. Their inability to mobilize universal moral principles is regarded as a reason for the rise of NS (Apel 1988). This historical argument can take a stronger form. Some authors argue that the relativism of post-Hegelian philosophy is a prerequisite of Nazi ideology. Here, Nazi ideology is classified as a radical kind of relativism that emerges from the general path of German philosophy after Hegel (Böhler 1988; Wolin 2004; Kellerwessel 2014).

The identification of NS with relativism has a long history and is still popular. The most influential account stems from Georg Lukács (1885–1971) who held a Hegelian Marxism when he published his polemical treatise *The Destruction of Reason* in 1954. The Neo-scholastic Josef de Vries (1898–1989) confronted Nazi philosophers with the charge of relativism already in the 1930s (de Vries 1935a, 1935b). Recently, proponents of discourse ethics combined their reading of NS as relativism with the warning that postmodernism represents a similar kind of relativism and could thus have devastating moral consequences (Apel 1988; Böhler 1988; Kellerwessel 2014).

The argument equating NS with relativism is therefore developed in varying contexts like Marxism, Catholicism, and discourse ethics. The different versions share, however, a common core: Most of these critical anti-relativists embed their identification of NS with relativism in a broader claim about the nature of philosophy. They argue that philosophy has to be based on reason and requires an orientation to a kind of absolute truth. Moreover, they defend the possibility of objective knowledge about reality and believe in a universal foundation of morality. These systematic convictions are usually connected with a claim about the historical development of post-Hegelian philosophy. The critical anti-relativists accuse especially historicism and *Lebensphilosophie* (philosophy of life) of having advocated a "dangerous" relativization of truth, knowledge, and values. On this view, the "relativistic nineteenth century" created a philosophical framework that enabled the flourishing of irrational beliefs, arbitrary maxims, and nihilistic attitudes. Ideologies such as NS are regarded as the ultimate step of the "destruction of reason" (Lukács 1954; see also de Vries 1935a, 1935b; Lieber 1966; Apel 1988; Böhler 1988; Wolin 2004).

Recent accounts highlight the destruction of moral rationality by the alleged relativism of Nazi racism. They read Nazi ideology as biological determinism that attributes mutually exclusive sets of values to the alleged races. The particular values of a race are chosen arbitrarily and are understood only instrumentally, since their realization should ensure the survival and flourishing of the respective race. The alleged racial hierarchy has no normative foundation and is thus completely arbitrary, too. This "extreme relativism" of NS is defined as the opposite to moral rationality and is considered as an attack against philosophy itself (Böhler 1988; Tugendhat 2009; Kellerwessel 2014).[1] Earlier accounts emphasize the opposition of NS to rationality in general. Lukács characterizes Nazi ideology as a modern myth that is nothing more than demagogic and nihilistic propaganda designed to deceive the population. Here, NS is portrayed as the consequence of the decay of philosophy that was caused by relativism. Following Lukács, Lieber (1923–2012) explicitly defines NS as the "end of philosophy" (Lieber 1966, 93).[2]

There are several reasons why the argument equating NS with relativism is problematic.

First, the argument rests on strong background assumptions about the nature of philosophy and morality. Most presentations of the argument equating NS with relativism take absolute standards for granted and thus lack a proper justification of their default position. This is problematic because the critical anti-relativists hold conflicting views such as Marxism, Catholicism, or discourse ethics: their versions of absolute truth, objective knowledge, and universal values contradict each other.

Second, the concept of relativism is used only in a pejorative sense. Relativism is often identified with the lack of what rationality consists in.

Third, the pejorative use of the concept of relativism makes the historical argument problematic. There are hardly any philosophers who actually held the kind of relativism that is presented by the critical anti-relativists. Their historical accounts are generally uncharitable and lack philological scrutiny. Representatives of historicism and *Lebensphilosophie* are used as mere whipping boys. The one-sided portrayals of Dilthey (1833–1911) or Nietzsche (1844–1900) as typical relativists ignore or misrepresent their actual engagement with the problem of relativism.[3]

Fourth, the argument equating NS with relativism rests on a poor understanding of Nazi ideology. Recent historical research shows that Nazi ideology can be reduced neither to deceitful propaganda nor to simple biologism. The mere fact that many professional philosophers contributed to Nazi ideology should already make us doubt the equation of National Socialism with "the end of philosophy." The critical anti-relativists invoke a normative notion of philosophy that does not correspond to the historical reality.

Fifth, and most importantly, most critical anti-relativists do not consider the actual debate about relativism in the context of NS. Since Nazi philosophers were accused of being relativists by their contemporaries such as de Vries (1935a; 1935b), they engaged seriously with the problem of relativism. The actual contributions of Nazi philosophers to this debate reveal their self-understanding and are therefore an important source for defining the relation of NS to relativism. The critical analysis of this engagement also shows us the meaning and the use of the concept of relativism in the historical context. In the next section, I examine this historical context.

The ideological debate: NS versus relativism

Nazi ideologues and philosophers were confronted with philosophical problems such as relativism because of the comprehensive character of their political claims. NS considered itself as a political revolution that realizes a new image of the human. Recent historical research confirms the self-understanding and contemporary perception of NS as a *weltanschauliche Bewegung* (ideological movement; e.g., Kroll 1998; Szeynmann 2013; Raphael 2014).[4] These nuanced approaches to the ideological dimension of NS suggest a new understanding of its structure and explain long-ignored phenomena, like the high degree of self-mobilization of German academia (Sluga 1993; Wolters 1999; Raphael 2014).

Nazi ideology has to be seen as a set of basic beliefs and convictions which offered much scope for interpretation. Although key concepts, like race, had to be accepted as guidelines of thinking and acting, different interpretations of such ideological core elements coexisted and competed even in the inner circle of Nazi leadership. Put shortly, since there was no unified and mandatory ideological system, the well-known policracy of Nazi government was accompanied by the polycentrism of Nazi ideology.[5] Nevertheless, it does not follow from this lack of a dogmatic version that Nazi ideology was nothing but a chimera. The "combination of fluidity and flexibility with a set of convictions and core arguments" (Raphael 2014, 74) shows, instead, that a political ideology works best as controlled plurality. While demanding a general appeal and specific direction, the Nazi worldview remained open to individual and contextualized interpretations. Take the example of the concept of race: once you had accepted its key role for understanding whatever phenomenon interested you, you could engage in the heated debate on its meaning and significance. The range that was developed in the ideological writings of political leaders reached from bluntly biological conceptions (e.g., Darré [1902–1946]) to metaphysical interpretations of race (e.g., Rosenberg [1893–1946)). Such obvious tensions were never removed and created the impression that NS was always in need of further explication. The crudity of Nazi ideology was a key reason for the intensive collaborations of scholars.

Philosophers in particular took up the task of elaborating, justifying, and explaining what NS truly is. There was a veritable quest for the officially accepted philosophy of NS in which representatives from most camps of German philosophy participated. Many German philosophers thus welcomed NS and attempted to show its philosophical significance. They put their philosophy into political service. The gesture of general agreement with the political change and the willingness to work in the direction of the leader (*dem Führer entgegenarbeiten*) were even more widespread.[6]

In the following, I concentrate on a specific philosophical interpretation of NS and its claim to realize a new concept of humanity. A number of philosophers welcomed NS because of its political break with the humanist tradition. Philosophers including Alfred Baeumler (1887–1968), Ernst Krieck

(1882–1947), and Erich Rothacker (1888–1965) defined their own task as establishing a new conception of humanity in the realm of theory. Thus, anthropology became a paradigmatic way to understand NS philosophically. This strand of Nazi philosophy was politically relevant because its key motifs were shared by an important representative in the inner circle of Nazi leadership: Alfred Rosenberg.[7] My examination focuses on the radical views of Rosenberg, Baeumler, and Krieck, who were often attacked because of their alleged relativism (e.g., de Vries 1935a, 1935b).

Their emphasis on the anthropological significance of NS was motivated by a strict rejection of universal concepts of humanity. Rosenberg, Baeumler, and Krieck argued that universalist doctrines provided only abstract accounts of human life that did not capture its actual reality. They concluded that such approaches were false and, moreover, often suggested that all universal concepts of humanity are deceitful fictions. Universalist claims were defined as purely ideological mechanisms that should hide the imperialist aspirations of certain actors on the world stage. From a Nazi perspective, universalists suggested that a certain way of life is the only way of life and thus threatened the identity of all other people. This line of thought was often combined with a critique of modern culture. Many Nazi philosophers believed that in the wake of modernity, many people, in adapting to Western culture, lost their particular identity. The humanist tradition was accused of hiding the fact that a particular Western form of being human claimed to be the only form of being human.[8]

The idea of an "endangered identity" was not only a major motif of the Nazi critique of modernity, but it was also the starting point of a specific political anthropology. The invocation "Remember who you are" (Baeumler 1934a, 6) was a key formula of Nazi ideology that also propagated a specific solution to the problem of identity: "Race always tells us what we are" (Baeumler 1943a, 93). Here, the "racial awakening" of NS was defined as a political response to the alleged identity crisis of humanity in modernity. From a Nazi perspective, identity always meant collective identity, and the latter was constituted by belonging to a community. Moreover, an individual belonged to a community by birth, and hence the identity of a person was a fixed property. Since both the body and the thought of an individual was shaped by descent from a particular group, belonging to this community became the essential and sole dimension of a person's identity. This sublation of individuality to community was a key motif of Nazi ideology in general.

Most Nazi philosophers were convinced that race is an essential property of humans that structures the world. They presented this "racial particularism" as the anthropological alternative to the "raceless universalism" of the humanist tradition.[9] The basic motifs of their racist anthropology were: From the "racial standpoint," communities are the sole agents of human life. Communities are defined as distinct entities with natural and historical components: *Blut und Boden* (blood and soil). Race usually represents the natural component and is tantamount to a fixed type that could not be changed. But the racial types have to be realized in history by *Zucht* (breeding) and *Kampf* (fighting). Most Nazi philosophers regard history as nothing but the struggle of races for their self-realization as particular communities, i.e., as *Völker*. Thus, a community has to assert itself against all forms of otherness to become and remain the *Volk* it is: it has to be itself physically as well as spiritually, inwardly as well as outwardly. Each race has a spiritual center that is expressed in the cultural systems of its communities, including morality, science, and philosophy. The establishment of an *artgerechte* (species-appropriate) culture is an essential part of the self-realization of a community. The distinction between communities comes in degrees: communities from the same racial type are akin to each other and may understand each other on a basic level. Some races are, however, totally alien to each other and hence lack any mutual understanding.

Baeumler and Krieck drew epistemological consequences from the *völkisch* particularism of their political anthropology. Their epistemologies revealed the relativistic tendency of their thinking most clearly. They considered the community as the only source of epistemic authority and rejected all aspirations to universality, objectivity, and absoluteness in the realm of knowledge. As Krieck put it: "For us there is only one truth—but it is only for us." (Krieck 1934a, 17; see also 1936a, 1 f.) The belief in the normative authority of community gave rise to a radically socialized and politicized concept of knowledge. Baeumler and Krieck emphasized the revolutionary character of their social and political epistemology. The dismissal of universal values, objective knowledge, and absolute truths as mere fictions was presented as a radical renewal of philosophy from a Nazi perspective. The key motifs of its epistemological core were:

a) *Gebundenheit* (Dependence): The basic conviction of Nazi epistemology was that all knowledge is dependent on the social and historical context in which it emerges. Even scientific knowledge is bound to the racial-*völkisch* community that discovers and preserves these insights under specific circumstances. Consider Krieck's concept of science: he defended the idea that there is a distinct "German science" (*deutsche Wissenschaft*) and regarded its claims as nothing but "the expression, the impact of the German character and essence" (Krieck 1938, 28).[10] This radical form of dependence reduced all kinds of knowledge to the racial-*völkisch* framework that was developed by Nazi anthropology.

b) *Begrenztheit* (Limitation): Krieck also emphasized that the insights of "German Science" can only be understood within the German community. Moreover, he defined the borders of a community as the limits of the validity of its claims (Krieck 1938, 126 f.). Krieck assumed that our knowledge is constrained by the racial-*völkisch* community we live in.

c) *Appropriateness*: The basic criterion for all knowledge claims is whether their content is *artgerecht* or *artfremd* (appropriate for or foreign to the species). Hence, the epistemic status of a belief is relative to the character of the community. This epistemic criterion demonstrates the ultimate authority in the realm of knowledge: The community itself provides the justificatory standard for the validity of beliefs.

d) *Gerichtetheit* (Tendency): For Nazi philosophers, there was no objective knowledge because no neutral epistemic perspective is available. Baeumler claimed that the "will to knowledge" is always led by current interests and has to have a political tendency. He argued that science without political tendency is idle business and deceit (Baeumler 1934b, 107, 111 f., 1934c, 154). Krieck's concept of "German Science" is an example of this radical perspectivism: If scientific claims result from the impact of the German character, they rest on a specific attitude. Krieck even regarded natural laws as mere interpretation of the world from a specific racial-*völkisch* perspective (Krieck 1938, 129). For him, all science is "political science" (Krieck 1936a, 1 ff.).

e) *Science as war:* Baeumler and Krieck understood science as a war of perspectives. They defined the scientist as a warrior whose activity contributes to the self-realization of his racial-*völkisch* community. This martial imagery was applied to all levels of scientific activity: inquiries were depicted as literal struggles with problems (Baeumler 1934b, 112). Epistemic claims have to fight for their validity, too. They have to conquer diverging claims within their own epistemic system and opposing claims from epistemic systems of other communities. These inner and outer science wars are part of the general struggle for the realization of a specific racial-*völkisch* community.

At first sight, this basic picture of Nazi anthropology and its epistemological consequences seems to confirm the assessment of the critical anti-relativists. Nazi ideology sounds like a radical form of relativism that is applied across the board. The relativistic tendency of NS is, however, only the first part of the story.[11] This is because Nazi philosophers rejected the label relativism fiercely. They regarded relativism as a fundamental problem of modern societies and presented NS as the long-overdue political solution to the problem. Take, e.g., Rosenberg, who characterized the idea of the "relativity of the universe" as an "illness of our time" that was overcome by the "organic truth of NS" (Rosenberg 1938, 694). This "illness" was often characterized as a heritage of the nineteenth century for which intellectual tendencies such as individualism, liberalism, historicism, and pragmatism were responsible.[12] Moreover, relativism was considered as a result of the *zersetzende* (decomposing) impact of the "Jewish spirit."[13] This dismissal raises important interpretative questions regarding the relation of NS and relativism. In the following, I argue that Nazi philosophers had historical, political, and systematic reasons to reject relativism. Their anti-relativistic convictions expose the weakness of the received view of NS as relativism.

In the late nineteenth and early twentieth centuries, relativism was often presented as a dangerous consequence of the modern spirit. The societal changes of modernity and the insights into the historical plurality of human life were conceptualized as a loss of certainty. Construed as dissolution of fixed values, these relativistic tendencies seemed to bring about anarchy and nihilism (Windelband 1884, 116 f.; Dilthey 1898). Note that this common usage of relativism could have an anti-Semitic connotation.[14] Nazi philosophers thus had historical and political reasons to consider relativism as a problem that has to be overcome. But there were also systematic reasons to reject relativism from a Nazi perspective. Note that Nazi ideology was a *racist* anthropology. Nazi philosophers believed in an objective hierarchy of races and attempted to justify their ranking. The conviction that there is a *Herrenrasse* (master race) and that its superiority can be demonstrated is the non-relativistic core of Nazi ideology. Nazi philosophers did not advocate tolerance of other ways of life or keep neutral when being confronted with different worldviews. They ranked the cultural systems of other communities without qualification and deduced a claim to power from their ranking. Moreover, the "Nordic race" was often characterized as the *only* one whose communities possess the "creative strength" to develop culture. Hence, all cultural goods, including morality, science, and philosophy, were defined as achievements of "Nordic" communities. Here, truth and objectivity entered the picture again because they were defined as distinct values of the "master race." Krieck claimed straightfor-wardly: "Slaves do not know truth." (Krieck 1938, 125) He did not think that "slaves" have their own truth. On Krieck's view, they simply lack the intellec-tual and moral capacities to develop true insights.[15]

These non-relativistic assumptions of Nazi racism reveal a fundamental tension: On the one hand, Nazi philosophers rejected universal aspirations and absolute claims. Their emphasis on the dependence of values, know-ledge, and truth on the "racial-*völkisch*" community is a radical form of rela-tivization. Hence, Nazi ideology is characterized by a relativistic tendency. On the other hand, Nazi philosophers rejected relativism as well. They were convinced that there is a hierarchy of races and believed in an objective justifi-cation of their ranking. By defining truth and objectivity as distinct values of the master race, their non-relativistic features should be saved.

Most Nazi philosophers were aware of this tension and thus argued that NS overcomes the opposition between relativism and absolutism. They claimed that the Nazi worldview is tantamount to a third way in philosophy that is neither absolutist nor relativistic. Their argumentative strategies always referred back to the alleged special character of their framework of relativiza-tion: the "racial-*völkisch*" community.[16] Nazi biologists argued that that race is an objective concept that can be researched scientifically. Here the reduc-tion of knowledge to the racial-*völkisch* framework could be regarded as non-relativistic because it revealed the natural foundation of all knowledge claims, i.e., the racial types as well as their ranking (see Danneberg 2013, 157–162). This racist version of a naturalistic epistemology remained an often-invoked promise: the results of the "racial science" were rather poor and did not meet the high expectations of scientists, philosophers, and politicians (see Koonz 2003, 190–220).

Nazi thinkers thus developed alternative strategies to justify their racism. In particular, cultural arguments to underpin the racist hierarchy became more and more important. The political anthropology of Rosenberg, Baeumler, and Krieck provided an evaluative framework that could be applied to the philosophical problem of relativism as well. According to Rosenberg, the property of race is tantamount to the essence of humans that distinguishes them fundamentally from the animal world. He used the term *Rassenseele* (race-soul) to signify the deep, spiritual unity of human groups that cannot be found in nature. Rosenberg thus regarded a specific disposition as the essence of humanity: the capacity to develop a collective identity. Yet, he did not think that all humans possess a "race-soul." Rosenberg was convinced that only the "Nordic race" enables its members to create particular communities and hence to develop *völkische Persönlichkeiten* (*völkisch* personalities) (Rosenberg 1938, 249). Here, selfhood became the most important criterion to assess the value of a community: the more a community knows, realizes, and expresses itself, the better it is. Thus, it is the relationship to themselves that constitutes the superiority of these communities. Rosenberg regarded this particularist disposition as a prerequisite of cultural development. He thought, again like many of his fellow Nazi philosophers, that, in the contemporary world, only the Germans are capable of the deliberate particularism that marks the peak of humanity.

On this version of Nazi ideology, particularism becomes the standard to assess the value of a community and its cultural systems. This evaluative standard holds for epistemology, too. From a Nazi perspective, epistemic systems can be ranked according to their expression of the "racial-*völkisch*" essence of the world: the more particularist an epistemic system is, the more it corresponds to true reality of humanity (e.g., Krieck 1937, 33 f., 1938, 130). Universal views, on the other hand, do not capture the racial order of the world at all and are thus "degenerate" ideas (e.g., Baeumler 1937d, 126 f.). This line of thought offers a rationale within Nazi anthropology to argue that their particularist epistemology is neither relativistic nor absolutist. Nazi philosophers could argue that the insight into the "racial-*völkisch*" relativity of knowledge claims constitutes an epistemic privilege. This view is superior because a specific community develops a view that holds for all other communities, too and, hence, is not relativistic. But since *not* all communities are capable of developing the insight into the relativity of all knowledge claims, this view is neither absolute nor universal. Even the understanding of epistemic particularism is restricted to certain groups of people. Yet, this argument restricts only the availability of epistemic particularism. The general validity and alleged superiority of this view remain in tension with the relativistic tendency of its actual content. Remember the key motifs of the epistemologies of Baeumler and Krieck, such as limitation or appropriateness.

To sum up, Nazi philosophers developed various strategies to present their approaches as overcoming the opposition between absolutism and relativism. While their critique of absolutism has strong relativistic tendencies, their rejection of relativism is based on the non-relativistic core of their racism. Both aspects are an essential part of the self-understanding of Nazi philosophers despite the inner tensions of this position. These tensions could never be removed entirely. When we look at the critique of absolutism, NS seems to be a radical kind of relativism. When we consider the racist core of Nazi ideology, we find strong non-relativistic assumptions. Thus, Nazi philosophers often simply claimed that NS makes the debate over absolutism and relativism redundant (e.g., Baeumler 1943b, 196 f.; Krieck 1935b).

Nevertheless, the ambition to overcome the opposition between absolutism and relativism connected the ideological discussion with broader philosophical debates. Moreover, the promise to solve the problem of relativism in a new way was a main reason why philosophers considered NS as a political option in the historical context. The argument equating NS with relativism is thus a main hindrance to critically examining the philosophical collaboration with the Nazi regime. I turn to this topic in the next section.

The academic debate: The prize question of the prussian academy of science

Relativism was also a much-discussed philosophical issue in the academic debates during NS. In 1936, the historical-philosophical class of the Prussian Academy of Sciences announced a prize question on the topic: "the inner reasons of philosophical relativism and the possibility of its overcoming."[17] It was launched by Nicolai Hartmann (1882–1950) and supported by his colleague Eduard Spranger (1882–1963). Both held chairs in philosophy at the University of Berlin. Hartmann was one of the leading German philosophers in the early twentieth century and developed a new approach to ontology. He engaged in neither a philosophical justification of NS nor direct political activities. Yet, Hartmann stayed loyal to the Nazi regime and participated in representative academic projects. He organized the philosophical contribution to the well-known *Aktion Ritterbusch,* which was meant as *Kriegseinsatz der Geisteswissenschafte*n (war deployment of humanities). Moreover, Hartmann was a member of the German Philosophical Association (DPG) since 1917. Founded by Bruno Bauch (1887–1942) after his break with the Kant Society (*Kant Gesellschaft*) in 1917, the DPG pursued a nationalist and racist agenda in philosophy.[18]

The prize question of 1936 was also a prestigious project, not least because of its politically explosive topic. Spranger later claimed that Hartmann proposed this topic because he, too, was concerned that the time was "already very corroded by the illness of relativism" (Spranger 1960, 442).[19] There were eight anonymized submissions. The prize was awarded to Eduard May (1905–1956) at the Leibniz Day of the Prussian Academy of Science in 1939. The committee also praised Johannes Thyssen's (1892–1968) study *Das Problem des Relativismus* (The Problem of Relativism). Thyssen was an associate professor at the University of Bonn, where he was also appointed as full professor in 1947. May habilitated with his prize-winning study *Am Abgrund des Relativismus* (At the Abyss of Relativism) at the University of Munich in 1942. He was originally a biologist and turned to philosophy under the influence of Hugo Dingler's (1881–1954) philosophy of science. During NS, he worked as a scientist in the concentration camp Dachau and later, because of his expertise in pest control, in the concentration camp Auschwitz. May was, however, never a member of a party organization of the NSDAP. He could establish himself in academic philosophy after 1945 and was appointed as full professor at the Free University Berlin in 1951 (see Hoyer 2005; Klee 2005, 398).

Since Hartmann was disappointed with the outcome of the prize question, he convinced his assistant Hermann Wein (1912–1981) to engage with the problem of relativism. Wein's habilitation *Das Problem des Relativismus* was published in the volume *Systematische Philosophie* (1942), alongside prominent representatives of contemporary German philosophy, such as Arnold Gehlen (1904–1976) and Rothacker. The edition of this volume was Hartmann's contribution to the *Aktion Ritterbusch*. Wein followed Hartmann from Berlin to Göttingen in 1945 and was appointed as *außerplanmäßiger* (extraordinary) professor after Hartmann's death in 1950. He published a second edition of his relativism study in 1950 and changed all politically incriminating passages. In later bibliographies, he mentioned only the edition of 1950 as publication of his habilitation. (For details on Wein, see Tilitzki 2002, 863–867.)

What did academic philosophers think about relativism? Let us start with the set-up of the prize question itself. The prize question introduces a broad concept of relativism and attributes problematic consequences to this position. Any view that relativizes truth is defined as philosophical relativism. Since relativism has a *zersetzend* (decomposing) impact on philosophy and science, its emergence is considered as an epistemological problem that has to be overcome. The prize question suggests that every solution to the problem of relativism presupposes an insight into its "inner reasons," which are specified as hidden presuppositions that are shared by all forms of relativism and that indicate problems of philosophy in itself. Thus, relativism is regarded as the opposite of philosophy proper (see Thyssen 1955, XIV f.).

How did academic philosophers respond to this question? May's prize-winning study (=a) and Wein's follow-up study (=b) show us the general tendency of the academic approaches and enable us to define the relation of the academic discussion to the ideological debates.

(a) May identified the problem of relativism with the problem of truth in itself. He introduced a correspondence theory of truth and claimed that common sense, the ordinary practice of the most advanced empirical sciences and proper metaphysics presuppose this concept of truth. Moreover, he argued that any inquiry has to acknowledge specific epistemic values, namely clarity and consistency, to be in accordance with the logical core of truth: the law of non-contradiction. May defined disagreement as both the most important source of contradiction and the motivation for relativism. The relativist holds, according to May, that the two parties of a genuine disagreement can be equally right. Because of this deliberate violation of the law of non-contradiction, relativism is in opposition to truth and, hence, a threat to any form of knowledge.

May also tells us a well-known story about the emergence of modern relativism: he believed that the rise of empiricism in the nineteenth century caused the decay of the *a priori* and, therefore, relativism emerged. He claimed that empirical experience is not sufficient to overcome disagreement, not even in the natural sciences. The pursuit of truth has to be guided by an *überempirischer aber dennoch wirklichkeitsverankerter Maßstab* (meta-empirical, but nevertheless reality-rooted standard; May 1941, 136). May thought: "*The only anchor in the chaos of experience* is ... *the apriori*" (154). His proposal for a solution to the problem of relativism was thus a new apriorism. He claimed that our *unmittelbares Erlebnis* (immediate lived experience) of mind-independent facts involves "proper apriorities" (268). May believed that our basic experience of sensory qualities such as colors has to be understood as *erlebnishaftes Erfassen apriorischer Begriffe* (experiential grasping of a priori concepts; 237). The exposition of this new apriorism remained, however, sketchy.

Note that May dedicated the conclusion of his study to the clarification of the ideological standing of his position. He admitted that he defended a core principle of rationalism: the absolute concept of truth. But he highlighted that his approach was distinct from any form of Enlightenment and, moreover, opposed empiricism and positivism. May also stated that his view could be seen as being in conflict with the philosophy of Krieck who was one of the fiercest rivals of his mentor Dingler. May emphasized that he agreed with Krieck's emphasis on the "*völkisch*-racial relativity" of knowledge acquisition (294). But he rejected any attempt to relativize the validity of truth and knowledge to specific entities. Here, May noted correctly that Krieck sought to preserve truth and objectivity as epistemic privileges of the master race. He thus concluded that his position was not in conflict with Krieck, although he proposed that Krieck should clarify his view on the problem of validity. This is because some pupils of Krieck repudiated any quest for absolute truth and objective knowledge. May emphasized that he disagreed only with such radical views (295–297).

(b) When we look at Wein's case, the connection between the academic discussion and the ideological debate becomes even clearer. Wein did not develop a philosophy of NS, although he was a member of the NSDAP since 1937 and worked for the so-called *Amt Rosenberg* (Rosenberg office), which was responsible for ideological surveillance and education. Baeumler was a referee of Wein's habilitation and criticized the study fiercely. He assumed that his approach to the problem of relativism was motivated by opportunism (see Baeumler 1942). This assessment suggests that Wein's contribution shows us what Nazi philosophers wanted to hear about relativism during NS.

Wein presented relativism as a comprehensive problem that had to be overcome in all areas of life. His study emphasized the "evil" sources and "terrifying" consequences of relativism: Wein characterized relativism as a "miscarriage of Enlightenment" (Wein 1942, 530) that flourished in the nineteenth century, especially because of historicism and liberalism (440 f.). The emergence of this "spiritual illness" caused a deep "crisis" that paralyzed the minds of intellectuals and lead to defeatism and nihilism (441–443, see also 457, 459, 495, 539). Wein claimed that the "Jewish spirit" had a natural affinity to "relativistic thinking" (439) and was thus tantamount to a *Geist-Entartung* (degeneration of the spirit; 539).

In contrast to this excessive rhetoric, Wein's systematic argument was rather simplistic. Wein argued that knowledge always consists of a relation between a subject and an object. Relativism follows when you overestimate the subjective part of knowledge and fail to capture the significance of the object. Absolutism follows when you only consider the objective part and neglect the subject. Both totalizing views are false because of their one-sidedness (435–37). Wein believed that the Nazi revolution has politically overcome both relativism and absolutism (541). Yet, he did not accept the existing philosophical solutions. Wein did not even mention the ideological proposals of Baeumler and Krieck, and he explicitly rejected the academic proposals of May and Thyssen. Wein's own proposal for overcoming both absolutism and relativism remained rather vague. He bombastically called for a new philosophy that considers both the ontological foundation and the anthropological reality of humanity. He emphasized time and again that, on the one hand, human existence is rooted in the *Sein der Welt überhaupt* (real being of the world), but on the other hand constitutes an independent *Seinsstruktur* (structure of being) in the world (523–526, 558 f.) Wein adopted Carl Schmitt's concept *Großraumordnung* (order of the greater territory) and characterized the human relationship to the world as *geistige Großraum-Haltung* (the spiritual attitude of the greater territory; 498). But his key concept remained as vague as his anthropological ontology in general.

Both cases suggest the same conclusion: the academic discussion stayed within the framework of the politically accepted discourse and its forms of mutual criticism. The academic contributors refer to Nazi philosophers frequently, use ideological motifs, and often define their relation to the ideological proposals for overcoming relativism. The critical reception of the praised studies of May and Thyssen shows that they were regarded as part of the public discourse. Their proposals were discussed as serious contributions to the general debate on relativism.[20] Later attempts to present these contributions as a hidden critique of the "Nazi relativism of race," as Thyssen does in the preface of the second edition of his study in 1947, are misleading (Thyssen 1947, V). Although Thyssen did not engage in a philosophical justification of NS, he did not directly criticize Nazi ideology either. Thyssen even admits in the same preface of 1947 that a superficial reader of his study could have concluded, against his alleged intentions, that it conforms to some strands of Nazi ideology. His contribution was read in exactly that way. This only supports my claim that there is a basic conformity between the academic and the ideological debate on relativism during NS. From both perspectives, relativism is considered as a fundamental problem that has to be overcome. Although the academic philosophers remain within the confines of established technical debates, they put forward similar approaches to the problem of relativism to their ideological counterparts. The problem of relativism represents a continuity between Nazi ideology and academic philosophy. This conclusion is supported by Spranger's retrospective reflection in 1960: He claims that the humanities at Humboldt University succumbed to NS because of their struggle with historicism and their decline into relativism. Nazi ideology presented itself as the ultimate foundation of knowledge claims and, hence, seemed to offer a solution to the problem of relativism (Spranger 1960, 441 f.). To sum up, the prize question was an opportunity to show agreement with a leitmotif of Nazi ideology within the bounds of academic philosophy. It is thus a case of philosophical collaboration.

Conclusion

This chapter started with a critical examination of the widespread reproach that relativism and NS are connected with each other historically as well as systematically. My investigation of the actual debate on relativism during NS revealed a rather different picture: Nazi philosophers were convinced that their position overcomes the opposition between absolutism and relativism. They developed argumentative strategies to present NS as a third way in philosophy but could not resolve the tensions between the relativistic tendency and the anti-relativistic assumptions of their view. Nevertheless, the general ambition to solve the problem of relativism made their ideology attractive to contemporary philosophers. Anti-relativist sentiments were a strong motivating factor for the philosophical collaboration with NS. Equating NS with relativism hence obscures an important feature of Nazi ideology that partly explains its widespread philosophical acceptance in the historical context.

My chapter also shows that there is a common meaning of relativism in the context of NS. Nazi philosophers, academic philosophers, and Nazi critics share a specific understanding of relativism: They all consider relativism as a fundamental problem that has to be overcome. Relativism is depicted as a vague threat that endangers not only philosophy proper, but society and life in general. It is always the same strands of nineteenth-century philosophy who are found guilty of having caused this problem and the subsequent "crisis" of the modern spirit. Moreover, relativism is only used in a polemical sense: relativists are always the others, the philosophical and/or political enemies. To label someone a relativist is almost tantamount to making him an enemy. This common politics of relativism can be summarized by adopting the title of Käte Friedemann's (1932) article on this topic: the specter of relativism (*Das Gespenst des Relativismus*) haunts the context of NS. To banish this specter is a prerequisite for a less prejudiced and more effective critique of Nazi ideology and its ties to the philosophy of the early twentieth century.

Acknowledgment

This work was supported by the European Research Council (ERC) under Grant 339382.

For critical comments and helpful suggestions, I am indebted to audiences in Vienna and Frankfurt a. M., as well as to my co-editors.

Notes

1 There is an intense debate about the question of whether Nazi ideology qualifies as a moral position at all. The examination of this complex issue is beyond the scope of this chapter.
2 All translations are mine.
3 For careful examinations of Nietzsche's and Dilthey's approach to relativism see Leiter, Kinzel in this volume, see also Steizinger 2017.
4 *Weltanschauung* (worldview) was a vague concept that connected philosophy and politics in the early twentieth century. The concept captured comprehensive theories about the world that were meant to guide human actions. There was no sharp distinction between philosophy and *Weltanschauung*: Some philosophers considered their philosophy as foundation of a certain worldview and emphasized the political significance of their theories. Others attempted to separate the technical debates of academic philosophy from the popular claims of *Weltanschauungsphilosophie* (philosophy of worldviews).

   The Nazis adopted the concept *Weltanschauung* from the beginning in order to highlight the comprehensive character of their movement. I call their *Weltanschauung* an ideology because of its thoroughgoing political nature. Nazi ideologues such as Rosenberg dedicated their whole thinking to develop the Nazi worldview. Nazi philosophers such as Baeumler, Krieck, or Rothacker used their philosophical theories to justify NS. Most of them explicitly developed a philosophy of National Socialism. When I speak of Nazi philosophers, I mean both the ideologues and the philosophers. Academic philosophers engaged, first and foremost, in technical debates and attempted to keep philosophical theory and political practice apart.
5 The Nazi government was characterized by various circles of power who competed with each other for political influence (Kroll 1998, 19 f.).
6 For examples, see Sluga 1993; Wolters 1999; Sandkühler 2009; Sieg 2013. The research on philosophy during National Socialism is focused on Martin Heidegger. Since his case is not exceptional from a historical point of view, I concentrate on lesser-known philosophers who collaborated with NS at least for some time.
7 For details about Rosenberg's role and views see Kroll 1998, Steizinger 2018.
8 For this line of thought, see, e.g., Baeumler 1934b, 92 f.; 1937b; 1937c; 1943c; Krieck 1937, 8 f.; 1938, 25–28; Rosenberg 1934, 8 f.; 1935; 1938; 639 ff.; 671 f.
9 See, e.g., Baeumler 1937d, 126 f., 1943c, 1943d, 96 f.; Rosenberg 1935, 1938, 33, 81 f., 84 f., 105, 106 f. 479 f., 482 f.; Krieck 1936b, 42–44, 1935a, 1937, 45–37, 1938, 119 f.

10 In German: *Deutsche Wissenschaft ist nichts anderes als Ausdruck, Auswirkung deutscher Art und Wesenheit.*

11 Böhnigk (2016) denies that National Socialism has a relativistic tendency and emphasizes its universal aspirations against the critical anti-relativists. He ignores, however, the Nazi critique of universalism and downplays their commitment to racial particularism. Moreover, Böhnigk identifies Nazi ideology with biological racism and thus misses its general character.

12 See, e.g., Rosenberg 1938, 694 f.; Baeumler 1943d, 27 f., 67 ff.; Bauch 1934/35, 43 f., 50 f., 52; Del Negro, 10–13; Krieck 1936a, 3, 7 f., 1938, 11.

13 See, e.g., Bauch 1934/35, 43 f., 53 f.; Del Negro 1942, 11–13, 42 f.; Krieck 1934a, 16 f. The popular *Meyers Dictionary* of 1942 defines relativism as the spiritual attitude that represents the modern liberal-individualistic perplexity and lack of principles. Jews occupy this position because of their disposition, and intentionally for the purpose to decompose society (*Meyers Lexikon* 1942: 290). Another example is the Nazi critique of Einstein's theory of relativity (see Herbert 2001, 12–14, 213; Danneberg 2013, 74 ff).

14 See, e.g., Köhnke (1996, 476–478), who shows the anti-Semitic connotations of the early reception of Simmel's relativism. Simmel is also the prime example of the decomposing relativism of the "Jewish spirit" in Del Negro's Nazi account of contemporary philosophy (Del Negro 1942, 11 f., 42 f.). For a careful account of Simmel's relativism and its diverse reception see Kusch, this volume.

15 For this line of thought see esp. Rosenberg 1938, chapter I, VI; see also, e.g., Baeumler 1934c, 1937a; Krieck 1936b, 60 f., 1937, 46–48, 1938, 130 f. For a detailed account of Baeumler's racist particularism see Steizinger 2016. Because of the racist core of their ideology, Nazi philosophers cannot be seen as defenders of the "plurality of Völker and cultures" (Tilitzki 2002, 29). Tilitzki's broad study contains much intriguing material but develops an untenable and unacceptable reading of Nazi ideology.

16 See, e.g., Baeumler 1942, 1943b, 197; Krieck 1936b, 16 f., 1935, 318 f., 1937, 33f, 1938, 11.

17 In German "Die inneren Gründe des philosophischen Relativismus und die Möglichkeit seiner Überwindung." Danneberg (2013) addresses the prize question in his extensive reconstruction of the attempts of Nazi scientists and philosophers to develop new standards of epistemic validity.

18 Schefczyk and Kuchinsky (2016) give a convincing account of Hartmann's ambivalent role during National Socialism. For the political orientation of the DPG, see Sluga 1995; Sieg 2013. For the significance of the *Aktion Ritterbusch,* see Hausmann 2007.

19 In German: *von der Krankheit des Relativismus schon stark angefressen.*

20 For a summary of the critical reception of May's and Thyssen's studies, see Danneberg 2013, 441 ff.

References

Apel, K.-O. (1988), "Zurück zur Normalität? Oder könnten wir aus der nationalen Katastrophe etwas Besonderes gelernt haben? Das Problem des (welt-) geschichtlichen Übergangs zur postkonventionellen Moral in spezifisch deutscher Sicht," in Kuhlmann (1988), 91–142.

Baeumler, A. (1934a), "Der Sinn des großen Krieges," in *Männerbund und Wissenschaft*, Berlin: Junker und Dünnhaupt, 1‒29.

—— (1934b), "Die geistesgeschichtliche Lage im Spiegel der Mathematik und Physik," in *Männerbund und Wissenschaft*, 75‒93.

—— (1934c), "Der theoretische und der politische Mensch," in *Männerbund und Wissenschaft* S. 94‒138.

—— (1934d), "Der politische Student," in *Männerbund und Wissenschaft*, 149‒156.

—— (1937a), "Rosenberg der Ghibelline," in *Politik und Erziehung*, Berlin: Junker und Dünnhaupt, 16‒28.

—— (1937b), "Der politische Volksbegriff," in *Politik und Erziehung*, 43‒49.

—— (1937c), "Der Kampf um den Humanismus," in *Politik und Erziehung*, 57‒66.

—— (1937d), "Kultur und Volk. Die Begründung der deutschen Leibesübungen," in *Politik und Erziehung*, 123–138.

—— (1942), Gutachten zur Habil.-Schrift Wein, Archive, Humboldt-University Berlin, Kur. W 93, PA Wein.

—— (1943a), "Nationalsozialismus und 'Idealismus'," in *Bildung und Gemeinschaft*, 2nd ed., 86‒97.

—— (1943b), "Philosophie," in *Bildung und Gemeinschaft* 196–198.

—— (1943c), *Weltdemokratie und Nationalsozialismus*, Berlin: Dunker & Humblot.

—— (1943d), *Alfred Rosenberg und der Mythos des 20. Jahrhunderts*, München: Hoheneichen.

Bauch, B. (1934/35), "Wert und Zweck," *Blätter für Deutsche Philosophie* VIII: 39–59.

Böhler, D. (1988), "Die deutsche Zerstörung des politisch-ethischen Universalismus. Über die Gefahr des – heute (post-) modernen – Relativismus und Dezisionismus," in Kuhlmann (1988), 166–216.

Böhnigk, V. (2016), "Eine Beziehung zwischen Relativismus und Nationalsozialismus – Tatsache oder Fiktion?" *Jahrbuch zur Geschichte und Wirkung des Holocaust* 20: 243–262.

Danneberg, L. (2013), *Wissenschaftsbegriff und epistemischer Relativismus im Nationalsozialismus*, url: http://fheh.org/wp-content/uploads/2016/07/relativismusld.pdf (18/03/16).

Del Negro, W. (1942), *Die Philosophie der Gegenwart in Deutschland*, Leipzig: Felix Meiner.

De Vries, J. (1935a), "Wissenschaft, Weltanschauung, Wahrheit," *Stimmen der Zeit* 129: 93–105.

—— 1935b, "Rationale oder irrationale Weltanschauung," *Stimmen der Zeit* 129: 380–392.

Dilthey, W. (1898), "Die Kultur der Gegenwart und die Philosophie," in *Gesammelte Schriften*, vol. VIII, 3rd ed., Göttingen: Vandenhoeck & Ruprecht, 1962, 190–205.

Friedemann, K. (1932), "Das Gespenst des Relativismus," *Philosophisches Jahrbuch* 45: 18–34.

Hausmann, F.-L. (2007) *"Deutsche Geisteswissenschaft" im Zweiten Weltkrieg: die "Aktion Ritterbusch" (1940–1945)*, 3rd ed., Heidelberg: Synchron.

Herbert, C. (2001), *Victorian Relativity: Radical Thought and Scientific Discovery*, Chicago: University of Chicago Press.

Kellerwessel, W. (2014), "Universalism and Moral Relativism: On Some Aspects of the Modern Debate on Ethics and Nazism," in, *Nazi Ideology and Ethics*, edited by W. Bialas and L. Fritze, Newcastle: Cambridge Scholars, 367–387.

Klee, E. (2005), *Das Personenlexikon zum Dritten Reich*, 2nd ed., Frankfurt a. M.: Fischer.

Köhnke, K.-Ch. (1996), *Der junge Simmel in Theoriebeziehungen und sozialen Bewegungen*, Frankfurt a. M.: Suhrkamp.

Koonz, C. (2003), *The Nazi Conscience,* Cambridge, Mass.: Harvard University Press.

Krieck, E. (1934), *Wissenschaft, Weltanschauung, Hochschule,* Leipzig: Armanen-Verlag.

—— (1936a), *Nationalpolitische Erziehung* (1932), 20th ed., Leipzig: Armanen-Verlag.

—— (1936b), *Völkisch-politische Anthropologie*, vol. 1, Leipzig: Armanen-Verlag.

—— (1935a), "Ist der Nationalsozialismus 'universalistisch'?" *Volk im Werden* 3: 184–186.

—— (1935b), "Jesuitischer Relativismus," *Volk im Werden* 3: 316–319.

—— (1937), *Völkisch-politische Anthropologie*, vol. 3, Leipzig: Armanen-Verlag.

—— (1938), *Völkisch-politische Anthropologie*, vol. 3, Leipzig: Armanen-Verlag.

Kroll, F.-L. (1998), *Utopie als Ideologie. Geschichtsdenken und politisches Handeln im Dritten Reich*, Paderborn: Schöningh.

Kuhlmann, W. (ed.) (1988), *Zerstörung des moralischen Selbstbewußtseins: Chance oder Gefährdung?* Frankfurt a. M.: Suhrkamp.

Lieber, H.-J. (1966), "Die deutsche Lebensphilosophie und ihre Folgen," in Freie Universität Berlin (ed.), *Nationalsozialismus und die deutsche Universität*, Berlin: De Gruyter, 92–108.

Lukács, G. (1954), *Die Zerstörung der Vernunft. Der Weg des Irrationalismus von Schelling zu Hitler*, Berlin: Aufbau.

May, E. (1941), *Am Abgrund des Relativismus*, Berlin: Lüttke.

Meyers Lexikon (1942), vol. 9, 8th ed., Leipzig: Bibliographisches Institut.

Raphael, L. (2014), "Pluralities of National Socialist Ideology. New Perspectives on the Production and Diffusion of National Socialist *Weltanschauung*," in *Visions of Community in Nazi Germany. Social Engineering and Private Lives*, edited by M. Steber and B. Gotto, Oxford: Oxford University Press, 73–86.

Rosenberg, A. (1934), *Krisis und Neubau Europas,* Berlin: Junker und Dünnhaupt.

—— (1935), "Die Weltanschauung in der Außenpolitik", in *Gestaltung der Idee. Blut und Ehre Bd. II*, 10th ed., München: Franz Eher, 1939, 246–262.

—— (1938), *Der Mythus des 20. Jahrhunderts* (1930), 125th–128th ed., München: Hoheneichen.

Sandkühler, H. J. (2009), *Philosophie im Nationalsozialismus*, Hamburg: Felix Meiner.

Sieg, U. (2013), *Geist und Gewalt. Deutsche Philosophen zwischen Kaiserreich und Nationalsozialismus*, München: Carl Hanser.

Spranger, E. (1960), "Das Historismusproblem an der Universität Berlin seit 1900," in *Gesammelte Schriften*, vol. 5, Tübingen: Max Niemeyer, 1969, 430–446.

Sluga, H. (2013), *Heidegger's Crisis: Philosophy and Politics in Nazi Germany*, Cambridge, Mass.: Harvard University Press.

Szeynmann, C.-Ch. (2013), "Nazi Economic Thought and Rhetoric During the Weimar Republic: Capitalism and its Discontents," *Politics, Religion & Ideology* 14: 355–376. –

Steizinger, J. (2016), "Politik versus Moral. Alfred Baeumlers Versuch einer philosophischen Interpretation des Nationalsozialismus," *Jahrbuch zur Geschichte und Wirkung des Holocaust*, vol. 20, 29–48.

—— (2017), "Reorientations of Philosophy in the Age of History: Nietzsche's Gesture of Radical Break and Dilthey's Traditionalism," *Studia Philosophica* 76: 223–244.

—— (2018), "The Significance of Dehumanization: Nazi Ideology and its Psychological Consequences," *Politics, Religion & Ideology* 19: 139–157.

Thyssen, J. (1941), (1947), (1955), *Der philosophische Relativismus*, Bonn: Ludwig Röhrscheid.

Tilitzki, Ch. (2002), *Die deutsche Universitätsphilosophie in der Weimarer Republik und im Dritten Reich*, Berlin: De Gruyter.

Tugendhat, E. (2009), "Der moralische Universalismus in Konfrontation mit der Nazi-Ideologie," *Jahrbuch zur Geschichte und Wirkung des Holocaust* 13: 61–75.

Wein, H. (1942), "Das Problem des Relativismus," in N. Hartmann (ed.) *Systematische Philosophie*, Stuttgart and Berlin: Kohlhammer, 431–559.

Windelband, W. (1884), "Kritische oder Genetische Methode," in *Präludien. Aufsätze und Reden zur Philosophie und ihrer Geschichte*, vol. 2, 5th ed.,Tübingen: Mohr, 1915, 99–135.

Wolin, R. (2004), *The Seduction of Unreason: The Intellectual Romance with Fascism from Nietzsche to Postmodernism*. Princeton: Princeton University Press.

Wolters, G. (1999), "Der 'Führer' und seine Denker. Zur Philosophie des 'Dritten Reichs,'" *Deutsche Zeitschrift für Philosophie* 47: 223–251.

Index

Sombart thesis 175
soul-culture 173
space and time 60, 61, 80, 81, 83, 84, 85,
    141, 167
special sciences 26, 27, 33, 35
species relativism *see* anthropologism
specific relativism *see* anthropologism
Spencer, Herbert 60
Spenglerism 177–178
Spengler, Oswald 172, 199
Spranger, Eduard 242
Stallo, John Bernard 83–84
Steizinger, Johannes 37
stereoscopic vision 67
Strauss, Leo 152, 161n3
subjectivation 121
subjectivity 45–46, 71, 94, 98, 107,
    118–119, 122, 154
subjectivization 46

tastes 147
Taylor, Charles 203
Thucydides 134–135, 139, 142, 144; and
    perspectivism 144–147
Thyssen, Johannes 242–243, 245
time and space 60, 61, 80, 81, 83, 84, 85,
    141, 167
Tracz, R. Brian 73
tradition 48, 49; continuity of 51, 53, 55,
    56n8; play-like interaction 51
transcendental absolutism 27–30, 34, 36
transcendental attitude 100
transcendental phenomenology 103–104;
    absolute validity of 107
transcendental theory 166
Troeltsch, Ernst 6, 7, 26, 38n1, 39n5,
    168, 173, 175; *Der Historismus und
    seine Probleme (Historicism and Its
    Problems)* 17
truth 155; critical theory of 155; and
    individualism 98
Tugendhat, Ernst 194n6

unified physical theory 163n25
unity-in-diversity 209–212
universalism 5, 237; immanent
    universalism 27–30, 35; and
    particularism 5; universal history 14
Urban, Wilbur Marshall 113, 114, 124n5

value judgments 119
values 44; general theory of 113–116;
    philosophy of 113–116, 131

visual experience 65, 75n4
Vogt, Karl 129
Volksgeist (national spirit) 5
Voltaire 19, 22n4
von Ehrenfels, Christian 114, 116

Walzer, Michael 212
Weber, Ernst 66, 67; *The Sense
    of Touch and the Common
    Sense* 66, 67
Weber–Fechner law 66, 116
Weber, Max 175
Weber's law 116
Wegener, Daniel 85
Wein, Hermann 243, 244–245
Weinstein, Deena 176
Weinstein, Michael A. 176
Weltanschauung see world-views
Weltgeist (world-spirit) 5
Weyl, Hermann 89
whole of experience/nature
    157–158
will 114
willing 141; and action 160
Winckelmann, Johann 203–204
Windelband, Wilhelm 6, 27,
    39n2, 39n3, 39n10, 117, 131;
    ahistorical philosophy 34–36;
    ahistorical relativism 36–38;
    brand of Neo-Kantianism 28;
    *Einleitung in die Philosophie
    (Introduction to Philosophy)* 33;
    judgments and evaluations 28; on
    philosophical problems 33–34;
    philosophy 31–34; transcendental
    absolutism and immanent
    universalism 27–29, 34
Wissenschaft 142
Wittgenstein, Ludwig 216
Wittkau, Annette 39n1
Wolters, Gereon 84
world history: general organization of
    14–15; individuals and analogies in
    16–17; totality of 12
world-views 7, 33, 35, 36, 37, 40n11, 95,
    101–102, 247n4
Wrathall, Mark A. 186, 189
Wundt, Wilhelm 60, 65, 116

Yorck, Ludwig 181

Ziehen, Theodor 116
Zuckert, Rachel 209